# Applied
# Math

# Applied Math

**Avi C. Bajpai**
*Loughborough University of Technology*

**Rodney M. Bond**
*Burleigh College, Loughborough*

*Adapted by*
**Jerry W. Jones**
Aiken Technical College
Aiken, South Carolina

PRENTICE HALL, Englewood Cliffs, New Jersey 07632

*Library of Congress Cataloging in Publication Data:*

Bajpai, A. C. (Avinash Chandra)
  Applied math

  Includes index.
  1. Mathematics—1961-          I. Bond, R. M.
II. Jones, Jerry W.          III. Title.
QA39.2.B354     1983          513'.14                    82-10936

Printed in the United States of America

10  9  8  7  6  5  4  3  2  1

ISBN   0-13-037938-7

Prentice-Hall International (UK) Limited, *London*
Prentice-Hall of Australia Pty. Limited, *Sydney*
Prentice-Hall Canada Inc., *Toronto*
Prentice-Hall Hispanoamericana, S.A., *Mexico*
Prentice-Hall of India Private Limited, *New Delhi*
Prentice-Hall of Japan, Inc., *Tokyo*
Simon & Schuster Asia Pte. Ltd., *Singapore*
Editora Prentice-Hall do Brasil, Ltda., *Rio de Janeiro*

*To Linda and Wayne*

# Preface

This book provides a thorough but concise coverage of the applied mathematics needed in many trade areas such as machine technology, drafting, automotive technology, sheet-metal fabrication, welding, air conditioning, landscaping, and many other occupations. It is intended for those students and instructors who want a direct, practical approach to applied mathematics, and it is suitable for a two-quarter or one-semester applied mathematics course for adults.

The philosophy of this textbook has been to teach by example, rather than by discussion only, and to reinforce math skills by the frequent use of practical applications, examples, and exercises.

The book has been organized into four major parts—arithmetic, algebra, geometry, and trigonometry—each consisting of several chapters. Each chapter contains frequent **Self-Assessment Tests** at logical points in the material, which may be used for self-paced courses or by the instructor to gauge students' progress. Each section of a chapter is accompanied by at least one set of **Practice Exercises** on the particular math skill under discussion. **Practical Applications Exercises** follow the coverage of most major topics. They are drawn from industrial experience, and many are similar to those questions used on professional and apprentice exams. In order to give the student frequent practice with applications, there are as many as four sets of **Practical Applications Exercises** in each chapter. They are appropriately placed to reinforce the skills just learned. At the end of each chapter are **Miscellaneous Exercises** which are similar to the **Practical Applications Exercises,** but draw on the skills of the entire chapter.

Three special appendixes are included on **Scientific Notation, Metrics,** and **The Calculator.** They are similar in structure to the chapters, but may be inserted in the course at any point or omitted if not needed.

The sections in which the calculator can be of use (see **Appendix C**) are referred to in the main chapters at each point where a new calculator skill might be needed. In this way the student can gradually build the skills required for use with the calculator. Tables are provided and covered in the text for those who do not use a calculator.

**Appendix B**, on metrics, may be covered as a unit or in sections. It concentrates on the common units of weight, length, area, volume, and capacity. The approach is on conversion within the metric system, not on English to metric conversions, based on the recommendations of the U.S. Metric Association and the various mathematics organizations.

About 40% of the problems and exercises in the text use metric measurements. Problems and exercises using metric measurements are noted (throughout the text) by an asterisk. Most of these do not require any background in metrics, other than familiarity with the unit names and abbreviations. Therefore, it is not necessary to cover **Appendix B** unless a more detailed knowledge is needed.

To maintain flexibility in the text discussions, all of the answers to the **Self-Assessment Tests** are included in **Appendix E** at the end of the book. The answers to the odd-numbered **Practice Exercises, Practical Applications Exercises,** and **Miscellaneous Exercises** are also included in the answer section.

For the convenience of the instructor, the Instructor's Guide includes all answers to exercises, as well as additional practical applications problems and sample chapter tests.

*A. C. Bajpai*
*R. M. Bond*
*J. W. Jones*

# Contents

# PART II   APPLIED ALGEBRA

# PART III   APPLIED GEOMETRY

# Applied
# Math

# PART I

## Applied Arithmetic

# 1
# Common Fractions

*Self-Assessment Test No. 1*

Convert the following mixed numbers to improper fractions.

(1)  $1\frac{1}{2}$  (2)  $2\frac{3}{4}$  (3)  $7\frac{1}{5}$  (4)  $8\frac{1}{8}$

(5)  $3\frac{1}{12}$  (6)  $1\frac{9}{16}$  (7)  $2\frac{1}{3}$  (8)  $9\frac{8}{9}$

Convert the following improper fractions to mixed numbers.

(9)  $\frac{13}{2}$  (10)  $\frac{15}{4}$  (11)  $\frac{17}{3}$  (12)  $\frac{10}{7}$

(13)  $\frac{24}{11}$  (14)  $\frac{80}{7}$  (15)  $\frac{5}{3}$  (16)  $\frac{49}{5}$

## 1.1  INTRODUCTION TO COMMON FRACTIONS

### Types of Fractions

**Proper Fractions (less than one unit)**

For example, $\frac{7}{8}, \frac{3}{5}, \frac{9}{16}$. The denominator (bottom number) is bigger than the numerator (top number).

**Improper Fractions (bigger than one unit)**

For example, $\frac{13}{5}, \frac{3}{2}, \frac{5}{4}$. The denominator is less than the numerator.

3

## Mixed Numbers

For example, $1\frac{3}{4}$, $7\frac{5}{8}$, $6\frac{1}{3}$. These consist of a whole number together with a proper fraction.

## Practice Exercise No. 1

Neatly copy Table 1–1 and check the correct column to show whether the figures in the first column are proper fractions, improper fractions, or mixed numbers.

## To Express a Mixed Number as an Improper Fraction

Consider $1\frac{2}{3}$. This means 1 whole unit $+\frac{2}{3}$ of a unit.

$$= 5 \text{ thirds} = \frac{5}{3}$$

1 whole unit     $\frac{2}{3}$ unit

Therefore, $1\frac{2}{3} = \frac{5}{3}$. (The symbol $\therefore$ is a shorthand for "therefore"; it is used frequently in this text.)

Consider $2\frac{3}{4}$. This means 2 whole units $+\frac{3}{4}$ of a unit.

$$\therefore 2\frac{3}{4} = \frac{11}{4}$$

$$= 11 \text{ quarters} = \frac{11}{4}$$

1 whole unit     1 whole unit     $\frac{3}{4}$ unit

## Practice Exercise No. 2

Using the above method, convert the following mixed numbers to improper fractions.

(1) $2\frac{1}{2}$     (2) $3\frac{1}{4}$     (3) $1\frac{1}{5}$     (4) $2\frac{1}{4}$     (5) $2\frac{4}{5}$

(6) $1\frac{7}{8}$     (7) $1\frac{3}{10}$     (8) $2\frac{1}{3}$     (9) $1\frac{7}{9}$     (10) $2\frac{1}{6}$

**TABLE 1–1**

|  | Proper Fraction | Improper Fraction | Mixed Number |  | Proper Fraction | Improper Fraction | Mixed Number |
|---|---|---|---|---|---|---|---|
| $\dfrac{7}{8}$ |  |  |  | $\dfrac{17}{3}$ |  |  |  |
| $1\dfrac{1}{5}$ |  |  | ✓ | $\dfrac{18}{37}$ |  |  |  |
| $\dfrac{7}{5}$ |  |  |  | $3\dfrac{9}{64}$ |  |  |  |
| $3\dfrac{1}{2}$ |  |  |  | $4\dfrac{1}{12}$ |  |  |  |
| $\dfrac{22}{7}$ |  |  |  | $\dfrac{17}{7}$ |  |  |  |
| $\dfrac{1}{4}$ |  |  |  | $\dfrac{114}{113}$ |  | ✓ |  |
| $\dfrac{9}{5}$ |  |  |  | $\dfrac{114}{115}$ |  |  |  |
| $\dfrac{16}{9}$ |  |  |  | $51\dfrac{2}{3}$ |  |  |  |
| $2\dfrac{4}{5}$ |  |  |  | $\dfrac{9}{75}$ |  |  |  |
| $\dfrac{11}{20}$ |  |  |  | $\dfrac{75}{9}$ |  |  |  |
| $\dfrac{3}{8}$ | ✓ |  |  | $3\dfrac{11}{15}$ |  |  |  |
| $7\dfrac{9}{11}$ |  |  |  | $\dfrac{14}{15}$ |  |  |  |
| $\dfrac{8}{5}$ |  |  |  | $\dfrac{173}{100}$ |  |  |  |
| $\dfrac{5}{32}$ |  |  |  | $\dfrac{16}{19}$ |  |  |  |
| $3\dfrac{9}{11}$ |  |  |  | $2\dfrac{3}{8}$ |  |  |  |

Drawing boxes is rather tedious. A simpler way is as follows.

To express a mixed number as an improper fraction, multiply the whole number by the denominator (bottom number) of the fractional part, add the numerator (top number) of the fraction to this product, and place the sum over the denominator. For example,

$$4\frac{2}{5} = \frac{(4 \times 5) + 2}{5} = \frac{22}{5}$$

$$7\frac{1}{3} = \frac{(7 \times 3) + 1}{3} = \frac{22}{3}$$

$$8\frac{1}{8} = \frac{(8 \times 8) + 1}{8} = \frac{65}{8}$$

### Practice Exercise No. 3

Copy and complete the following.

(1) $1\frac{4}{5} = \frac{(\quad \times \quad) +}{5} =$

(2) $3\frac{1}{8} = \frac{(\quad \times \quad) +}{8} =$

(3) $9\frac{1}{3} = \frac{(\quad \times \quad) +}{3} =$

(4) $1\frac{4}{15} = \frac{(\quad \times \quad) +}{15} =$

Sometimes we leave out the written work, calculating it mentally and write

$$5\frac{3}{4} = \frac{23}{4}$$

Now try to convert the following mixed numbers to improper fractions without showing the work.

(5) $3\frac{2}{3}$  (6) $3\frac{4}{5}$  (7) $8\frac{5}{6}$  (8) $2\frac{1}{9}$  (9) $3\frac{1}{3}$

(10) $7\frac{1}{16}$  (11) $1\frac{13}{15}$  (12) $4\frac{3}{5}$  (13) $2\frac{2}{3}$  (14) $8\frac{7}{10}$

(15) $11\frac{5}{8}$  (16) $4\frac{4}{9}$  (17) $10\frac{1}{10}$  (18) $6\frac{7}{10}$  (19) $5\frac{7}{12}$

(20) $19\frac{1}{2}$

## To Express an Improper
## Fraction as a Mixed Number

For example, express $\dfrac{9}{4}$ as a mixed number.

$\dfrac{9}{4}$ means 9 quarters.

(Remember 4 quarters make a whole unit.)

$$\therefore \frac{9}{4} = 2\frac{1}{4}$$

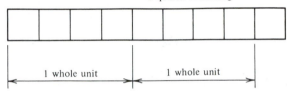

1 quarter remaining

1 whole unit      1 whole unit

When converting improper fractions to mixed numbers, divide the numerator (top number) by the denominator (bottom number) to obtain the whole number. The remainder, if any, when placed over the denominator gives the fractional part of the mixed number. For example,

$$\frac{21}{5} = 4\frac{1}{5} \qquad \frac{31}{7} = 4\frac{3}{7}$$

$$\frac{18}{9} = 2 \qquad \frac{33}{10} = 3\frac{3}{10}$$

## Practice Exercise No. 4

Express the following improper fractions as mixed numbers.

(1) $\dfrac{3}{2}$     (2) $\dfrac{7}{3}$     (3) $\dfrac{9}{4}$     (4) $\dfrac{8}{5}$     (5) $\dfrac{6}{5}$

(6) $\dfrac{4}{3}$     (7) $\dfrac{19}{4}$     (8) $\dfrac{32}{9}$     (9) $\dfrac{63}{10}$     (10) $\dfrac{81}{11}$

(11) $\dfrac{55}{8}$     (12) $\dfrac{19}{8}$     (13) $\dfrac{17}{4}$     (14) $\dfrac{16}{9}$     (15) $\dfrac{8}{7}$

(16) $\dfrac{9}{2}$     (17) $\dfrac{14}{5}$     (18) $\dfrac{17}{7}$     (19) $\dfrac{45}{9}$     (20) $\dfrac{63}{8}$

---

*Self-Assessment Test No. 2*

Copy and complete the following.

(1) $\dfrac{3}{4} = \dfrac{}{8}$

(2) $\dfrac{5}{6} = \dfrac{15}{}$

(3) $\dfrac{2}{5} = \dfrac{4}{} = \dfrac{}{25}$

(4) $\dfrac{7}{8} = \dfrac{49}{}$

(5) $\dfrac{4}{5} = \dfrac{24}{}$

(6) $\dfrac{21}{32} = \dfrac{42}{}$

Simplify the following fractions.

(7) $\dfrac{16}{20}$

(8) $\dfrac{21}{28}$

(9) $\dfrac{14}{28}$

(10) $\dfrac{12}{33}$

(11) $\dfrac{16}{64}$

(12) $\dfrac{21}{27}$

---

## To Change the Denominator of a Fraction

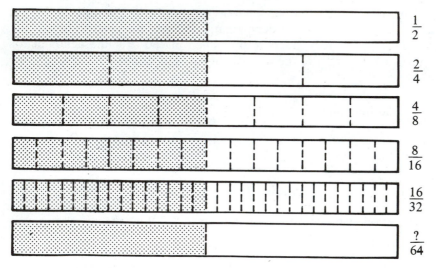

| | |
|---|---|
| | $\dfrac{1}{2}$ |
| | $\dfrac{2}{4}$ |
| | $\dfrac{4}{8}$ |
| | $\dfrac{8}{16}$ |
| | $\dfrac{16}{32}$ |
| | $\dfrac{?}{64}$ |

*Question* What can you say about the shaded areas above?

*Answer* They are the same.

$$\therefore \frac{1}{2} = \frac{2}{4} = \frac{4}{8} = \frac{8}{16} = \frac{16}{32} = \frac{32}{64}$$

A fraction is unaltered in value if the numerator and denominator are multiplied by the same nonzero number. This is equivalent to multiplying the complete fraction by 1. For example,

$$\frac{1}{5} = \frac{3}{15} \text{ (multiplying top and bottom by 3)}$$

$$\frac{1}{5} = \frac{5}{25} \text{ (multiplying top and bottom by 5)}$$

## Practice Exercise No. 5

Copy and complete the following:

(1) $\dfrac{1}{2} = \dfrac{2}{} = \dfrac{}{8}$

(2) $\dfrac{1}{4} = \dfrac{}{12} = \dfrac{5}{}$

(3) $\dfrac{3}{4} = \dfrac{}{8} = \dfrac{15}{}$

(4) $\dfrac{2}{3} = \dfrac{}{18} = \dfrac{10}{}$

(5) $\dfrac{4}{5} = \dfrac{20}{} = \dfrac{}{50}$

(6) $\dfrac{3}{8} = \dfrac{27}{} = \dfrac{}{64}$

(7) $\dfrac{6}{7} = \dfrac{24}{} = \dfrac{}{35}$

(8) $\dfrac{7}{10} = \dfrac{70}{} = \dfrac{}{60}$

(9) $\dfrac{1}{3} = \dfrac{3}{} = \dfrac{}{81}$

(10) $\dfrac{3}{4} = \dfrac{}{16} = \dfrac{27}{}$

## To Simplify a Fraction
## (Reduce to Lowest Terms)

Similar to the last section, the value of the fraction is unaltered if both numerator and denominator are divided by the same nonzero number. For example,

$$\frac{5}{20} = \frac{1}{4} \text{ (dividing top and bottom by 5)}$$

$$\frac{3}{27} = \frac{1}{9} \text{ (dividing top and bottom by 3)}$$

This process is often called "canceling" and is written as:

$$\frac{\overset{7}{\cancel{42}}}{\underset{10}{\cancel{60}}} = \frac{7}{10}$$

If reducing results in a 1 on the bottom, you have a whole number. For example,

$$\frac{30}{6} = \frac{5}{1} = 5$$

## Practice Exercise No. 6

Copy and complete the following.

**(1)** $\dfrac{16}{24} = \dfrac{}{12} = \dfrac{2}{}$   **(2)** $\dfrac{7}{21} = \dfrac{1}{}$   **(3)** $\dfrac{4}{16} = \dfrac{1}{}$

**(4)** $\dfrac{8}{56} = \dfrac{4}{} = \dfrac{1}{}$   **(5)** $\dfrac{8}{32} = \dfrac{2}{} = \dfrac{}{4}$   **(6)** $\dfrac{20}{64} = \dfrac{5}{}$

**(7)** $\dfrac{14}{2}$   **(8)** $\dfrac{24}{3}$   **(9)** $\dfrac{13}{13}$

Reduce the following fractions to lowest terms.

**(10)** $\dfrac{8}{24}$   **(11)** $\dfrac{8}{40}$   **(12)** $\dfrac{12}{15}$   **(13)** $\dfrac{9}{27}$

**(14)** $\dfrac{12}{60}$   **(15)** $\dfrac{18}{46}$   **(16)** $\dfrac{13}{52}$   **(17)** $\dfrac{10}{12}$

**(18)** $\dfrac{14}{42}$   **(19)** $\dfrac{18}{72}$   **(20)** $\dfrac{86}{144}$   **(21)** $\dfrac{18}{64}$

**(22)** $\dfrac{49}{84}$   **(23)** $\dfrac{72}{144}$   **(24)** $\dfrac{132}{156}$   **(25)** $\dfrac{81}{90}$

**(26)** $\dfrac{216}{432}$

---

*Self-Assessment Test No. 3*

Work out the following.

**(1)** $\dfrac{1}{3} + \dfrac{1}{8}$   **(2)** $\dfrac{1}{16} + \dfrac{1}{2} - \dfrac{1}{32}$

**(3)** $1\dfrac{1}{4} + 3\dfrac{1}{5}$   **(4)** $1\dfrac{2}{3} + 1\dfrac{1}{5}$

**(5)** $3\dfrac{1}{2} - 2\dfrac{5}{16}$   **(6)** $2\dfrac{1}{4} + 3\dfrac{1}{2} - 1\dfrac{1}{3}$

**(7)** $7\dfrac{1}{5} - 3\dfrac{1}{4} + 2$   **(8)** $6\dfrac{1}{8} + 2\dfrac{1}{16} - 3\dfrac{13}{32}$

---

## 1.2   ADDITION AND SUBTRACTION

### Addition and Subtraction of Proper Fractions

Before adding or subtracting fractions, they must be converted so that they have a common denominator—that is, numbers on the bottom of all fractions involved are the same. For example,

$$\frac{5}{8} + \frac{5}{16} = \frac{10}{16} + \frac{5}{16} = \frac{15}{16}$$

*Example (1)* $\quad \frac{2}{3} + \frac{5}{12} = \frac{8}{12} + \frac{5}{12} = \frac{13}{12} = 1\frac{1}{12}$

$$\left( \text{For convenience, this may be written } \frac{2}{3} + \frac{5}{12} = \frac{8 + 5}{12} = \frac{13}{12} = 1\frac{1}{12}. \right)$$

*Example (2)* $\quad \frac{1}{5} - \frac{1}{6} = \frac{6 - 5}{30} = \frac{1}{30}$

*Example (3)* $\quad \frac{1}{3} + \frac{1}{4} + \frac{5}{8} = \frac{8 + 6 + 15}{24} = \frac{29}{24} = 1\frac{5}{24}$

## Addition and Subtraction of Mixed Numbers

When adding or subtracting mixed numbers, the whole numbers and the fractions are dealt with separately. For example,

$$3\frac{1}{5} - 1\frac{1}{8}$$

Considering the whole numbers first, $3 - 1 = 2$.

The expression is now $2 + \frac{1}{5} - \frac{1}{8} = 2 + \frac{8 - 5}{40} = 2\frac{3}{40}.$

*Example (1)* $\quad 2\frac{1}{2} + 3\frac{5}{8} = 5 + \frac{1}{2} + \frac{5}{8}$

$$= 5 + \frac{4 + 5}{8} = 5 + \frac{9}{8}$$

$$= 5 + 1\frac{1}{8} = 6\frac{1}{8}$$

*Example (2)* $\quad 2\frac{1}{4} + 5\frac{1}{2} - 1\frac{3}{8} = 6\frac{2 + 4 - 3}{8} = 6\frac{3}{8}$

*Example (3)* $\quad 2\frac{1}{4} + 5\frac{3}{8} - 1\frac{7}{8} = 6\frac{2 + 3 - 7}{8} = 6\frac{5 - 7}{8}$

In this case, we cannot take 7 from 5, so 1 is taken from the whole number part, that is, from the 6 and converted to eighths. Therefore,

$$6\frac{5 - 7}{8} = 5 + \frac{8}{8} + \frac{5 - 7}{8} = 5\frac{8 + 5 - 7}{8} = 5\frac{6}{8} = 5\frac{3}{4}$$

## Practice Exercise No. 7

Work out the following.

(1) $\dfrac{1}{2} + \dfrac{1}{4}$

(2) $\dfrac{1}{8} + \dfrac{1}{16}$

(3) $\dfrac{2}{3} + \dfrac{3}{5}$

(4) $\dfrac{1}{2} + \dfrac{9}{16}$

(5) $\dfrac{1}{2} + \dfrac{3}{8} + \dfrac{7}{8}$

(6) $\dfrac{3}{8} + \dfrac{3}{4} + \dfrac{1}{8}$

(7) $\dfrac{3}{7} + \dfrac{5}{14}$

(8) $\dfrac{8}{9} + \dfrac{7}{8} + \dfrac{31}{72}$

(9) $\dfrac{3}{5} + \dfrac{8}{15} + \dfrac{7}{30}$

(10) $\dfrac{11}{12} + \dfrac{7}{24} + \dfrac{5}{6}$

(11) $\dfrac{1}{3} - \dfrac{1}{4}$

(12) $\dfrac{3}{8} - \dfrac{3}{32}$

(13) $\dfrac{1}{3} - \dfrac{2}{15}$

(14) $\dfrac{7}{16} - \dfrac{3}{32}$

(15) $\dfrac{8}{9} - \dfrac{13}{27}$

(16) $1\dfrac{1}{4} + 1\dfrac{3}{8}$

(17) $3\dfrac{8}{9} + 2\dfrac{1}{3}$

(18) $7\dfrac{1}{2} - 6\dfrac{1}{3}$

(19) $4\dfrac{4}{5} - 3\dfrac{7}{10}$

(20) $9\dfrac{1}{2} - 8\dfrac{1}{10}$

(21) $3\dfrac{1}{4} + 2\dfrac{2}{5} - 1\dfrac{9}{20}$

(22) $6\dfrac{5}{16} + 2\dfrac{3}{32} - 1\dfrac{1}{8}$

(23) $6\dfrac{1}{4} + 2\dfrac{1}{10} - 3\dfrac{1}{20}$

(24) $\dfrac{7}{64} - \dfrac{63}{64} + 1\dfrac{1}{2}$

(25) $1\dfrac{9}{11} - \dfrac{19}{22} + 1\dfrac{5}{11}$

(26) $1\dfrac{3}{8} - \dfrac{9}{16}$

(27) $2\dfrac{31}{32} + 1\dfrac{1}{16} - \dfrac{11}{16}$

(28) $3\dfrac{1}{5} + 2\dfrac{1}{4} - \dfrac{19}{20}$

(29) $6\dfrac{3}{5} + 7\dfrac{1}{8} - 3\dfrac{1}{10}$

(30) $8\dfrac{1}{8} - 6\dfrac{1}{4} + 2\dfrac{3}{32}$

---

*Self-Assessment Test No. 4*

Work out the following.

(1) $\dfrac{1}{3} \times \dfrac{1}{4}$    (2) $\dfrac{5}{8} \times \dfrac{7}{10}$    (3) $\dfrac{16}{25} \times \dfrac{5}{8}$

(4) $\dfrac{1}{4}$ of $\dfrac{1}{4}$    (5) $1\dfrac{1}{2} \times 1\dfrac{1}{3}$    (6) $2\dfrac{1}{2} \times 3\dfrac{1}{2}$

(7) $1\dfrac{1}{5} \times \dfrac{5}{6}$    (8) $\dfrac{1}{2} \div \dfrac{1}{4}$    (9) $\dfrac{1}{3} \div \dfrac{1}{3}$

(10) $\dfrac{5}{6} \div \dfrac{2}{5}$    (11) $1\dfrac{1}{3} \div \dfrac{2}{3}$    (12) $1\dfrac{5}{8} \div 4$

(13) $17\dfrac{1}{2} \div 3\dfrac{1}{2}$    (14) $1\dfrac{13}{32} \div 1\dfrac{1}{8}$

---

## 1.3   MULTIPLICATION OF FRACTIONS

   Shows $\dfrac{1}{2}$ of $\dfrac{1}{2}$ of a square or $\dfrac{1}{4}$ square.

   Shows $\dfrac{1}{3}$ of $\dfrac{1}{3}$ of a square or $\dfrac{1}{9}$ square.

  Shows $\dfrac{1}{4}$ of $\dfrac{1}{4}$ of a square or $\dfrac{1}{16}$ square.

$\dfrac{1}{2}$ of a $\dfrac{1}{2}$ means $\dfrac{1}{2} \times \dfrac{1}{2} = \dfrac{1}{4}$

$\dfrac{1}{3}$ of a $\dfrac{1}{3}$ means $\dfrac{1}{3} \times \dfrac{1}{3} = \dfrac{1}{9}$

$\dfrac{1}{4}$ of a $\dfrac{1}{4}$ means $\dfrac{1}{4} \times \dfrac{1}{4} = \dfrac{1}{16}$

When multiplying fractions, multiply numerator by numerator and denominator by denominator. For example,

$$\dfrac{3}{4} \times \dfrac{1}{5} = \dfrac{3 \times 1}{4 \times 5} = \dfrac{3}{20}$$

$$\dfrac{2}{3} \times \dfrac{1}{11} = \dfrac{2 \times 1}{3 \times 11} = \dfrac{2}{33}$$

Before multiplication, it is usually worthwhile to cancel wherever possible, because this will reduce the numbers in size and make any further work much easier.

$$\frac{1}{1}\frac{\cancel{4}}{\cancel{9}} \times \frac{\cancel{27}^{3}}{\cancel{32}_{8}} = \frac{3}{8}$$

$$\frac{1}{5}\frac{\cancel{2}}{5} \times \frac{7}{\cancel{8}_{4}} = \frac{7}{20}$$

**Note:** *Canceling can only take place when all numbers in the numerator are to be multiplied together, and all numbers in the denominator are to be multiplied together. Remember that "cancel" means divide the "top" and the "bottom" by the same number.*

## Practice Exercise No. 8

Work out the following.

(1) $\frac{1}{2} \times \frac{1}{4}$     (2) $\frac{3}{8} \times \frac{4}{5}$     (3) $\frac{4}{9} \times \frac{1}{5}$

(4) $\frac{2}{3} \times \frac{3}{4}$     (5) $\frac{6}{7} \times \frac{35}{36}$     (6) $\frac{7}{8} \times \frac{8}{49}$

(7) $\frac{8}{9} \times \frac{27}{32}$     (8) $\frac{1}{3} \times \frac{6}{7}$     (9) $\frac{3}{5} \times \frac{20}{33}$

(10) $\frac{8}{15} \times \frac{50}{64}$     (11) $\frac{1}{3}$ of $\frac{6}{11}$     (12) $\frac{1}{3}$ of $\frac{1}{8}$

(13) $\frac{1}{5}$ of $\frac{15}{32}$     (14) $\frac{7}{8}$ of $\frac{1}{4}$     (15) $\frac{4}{5}$ of $\frac{35}{64}$

## To Multiply a Fraction by a Whole Number

For example,

$$\frac{2}{3} \times 5 = \frac{2}{3} \times \frac{5}{1} = \frac{10}{3} = 3\frac{1}{3}$$

(Remember that $\frac{5}{1}$ is the whole number 5.)

$$\frac{7}{8} \times 10 = \frac{7}{4\cancel{8}} \times \frac{\cancel{10}^{5}}{1} = \frac{35}{4} = 8\frac{3}{4}$$

$$\frac{5}{8} \text{ of } \$800 = \frac{5}{\cancel{8}_{1}} \times \frac{\cancel{\$800}^{100}}{1} = \$500$$

## Practice Exercise No. 9

Work out the following.[1]

**(1)** $\frac{7}{8} \times 12$

**(2)** $\frac{3}{5} \times 85$

**\*(3)** $\frac{9}{16}$ of 320 mm

**(4)** $\frac{2}{25}$ of $300

**(5)** $\frac{6}{7} \times 42$

**(6)** $\frac{7}{10}$ of $990

**(7)** $\frac{8}{15}$ of $750

**(8)** $\frac{2}{3}$ of 345 ft

**(9)** $\frac{3}{4}$ of 154 lb

**\*(10)** $\frac{11}{12}$ of 72 g

**(11)** $\frac{5}{8}$ of $5.60

**(12)** $\frac{5}{12}$ of $6012

## Multiplication of Mixed Numbers

When multiplication involves mixed numbers, *always* change the mixed numbers to improper fractions and then multiply as before.

$$2\frac{1}{4} \times 1\frac{1}{8} = \frac{9}{4} \times \frac{9}{8} = \frac{81}{32} = 2\frac{17}{32}$$

$$1\frac{1}{5} \times \frac{15}{36} = \frac{\cancel{6}^1}{\cancel{5}_1} \times \frac{\cancel{15}^{\cancel{6}1}}{\cancel{36}_{\cancel{6}2}} = \frac{1}{2}$$

$$1\frac{3}{4} \times \frac{4}{15} = \frac{7}{\cancel{4}}1 \times \frac{\cancel{4}^1}{15} = \frac{7}{15}$$

## Practice Exercise No. 10

Work out the following.

**(1)** $1\frac{3}{4} \times 1\frac{1}{2}$

**(2)** $\frac{3}{3} \times 1\frac{4}{5}$

**(3)** $1\frac{3}{8} \times \frac{8}{11}$

**(4)** $7\frac{1}{2} \times 1\frac{1}{15}$

**(5)** $\frac{9}{32} \times 4\frac{4}{7}$

**(6)** $1\frac{1}{4} \times 2\frac{3}{5}$

**(7)** $\frac{3}{4} \times 1\frac{3}{5}$

**(8)** $1\frac{2}{3} \times \frac{1}{8}$

**(9)** $\frac{1}{4} \times 1\frac{7}{8}$

**(10)** $\frac{5}{12}$ of $4\frac{4}{5}$

**(11)** $\frac{1}{9}$ of $5\frac{1}{7}$

**(12)** $\frac{1}{10}$ of $3\frac{1}{3}$

**\*(13)** $\frac{1}{8}$ of 6000 m

**(14)** $\frac{2}{3}$ of 700 lb

**(15)** $\frac{13}{16}$ of 640 ft

---

[1] The asterisk (*) is used to indicate those problems using metric measurements.

## 1.4   DIVISION OF FRACTIONS

For example,

$$\frac{1}{4} \div \frac{2}{3}$$

When dividing by a fraction, invert (turn upside down) the fraction that you are dividing by and multiply so that

$$\frac{1}{4} \div \frac{2}{3} = \frac{1}{4} \times \frac{3}{2} = \frac{3}{8}$$

$$\frac{3}{4} \div \frac{5}{6} = \frac{3}{\cancel{4}_2} \times \frac{\cancel{6}^3}{5} = \frac{9}{10}$$

(*Never* cancel until after inverting.)

$$\frac{7}{8} \div \frac{5}{16} = \frac{7}{\cancel{8}_1} \times \frac{\cancel{16}^2}{5} = \frac{14}{5} = 2\frac{4}{5}$$

### Practice Exercise No. 11

Work out the following.

(1) $\dfrac{1}{8} \div \dfrac{1}{4}$     (2) $\dfrac{1}{3} \div \dfrac{1}{9}$     (3) $\dfrac{5}{7} \div \dfrac{2}{3}$     (4) $\dfrac{4}{5} \div \dfrac{3}{20}$

(5) $\dfrac{1}{4} \div \dfrac{1}{3}$     (6) $\dfrac{7}{10} \div \dfrac{7}{30}$     (7) $\dfrac{5}{4} \div \dfrac{3}{4}$     (8) $\dfrac{1}{16} \div \dfrac{3}{4}$

(9) $\dfrac{7}{32} \div \dfrac{1}{2}$     (10) $\dfrac{8}{11} \div \dfrac{5}{33}$     (11) $\dfrac{7}{12} \div \dfrac{4}{5}$     (12) $\dfrac{6}{7} \div \dfrac{4}{21}$

(13) $\dfrac{15}{16} \div \dfrac{3}{4}$     (14) $\dfrac{1}{2} \div \dfrac{1}{3}$     (15) $\dfrac{7}{8} \div \dfrac{21}{32}$     (16) $\dfrac{8}{9} \div \dfrac{64}{81}$

### Dividing a Fraction by a Whole Number

For example,

$$\frac{1}{2} \div 5$$

This requires writing 5 as $\dfrac{5}{1}$.

$$\therefore \frac{1}{2} \div 5 = \frac{1}{2} \div \frac{5}{1} = \frac{1}{2} \times \frac{1}{5} = \frac{1}{10}$$

$$\frac{1}{3} \div 10 = \frac{1}{3} \div \frac{10}{1} = \frac{1}{3} \times \frac{1}{10} = \frac{1}{30}$$

### Division Involving Mixed Numbers

When division involves mixed numbers, *always* change the mixed numbers to improper fractions and then divide as before.

$$1\frac{3}{8} \div 1\frac{1}{4} = \frac{11}{8} \div \frac{5}{4} = \frac{11}{\cancel{8}_2} \times \frac{\cancel{4}^1}{5} = \frac{11}{10} = 1\frac{1}{10}$$

$$1\frac{1}{4} \div 2\frac{3}{8} = \frac{5}{4} \div \frac{19}{8} = \frac{5}{\cancel{4}_1} \times \frac{\cancel{8}^2}{19} = \frac{10}{19}$$

Notice the order:

1. Change to improper fractions.
2. Invert the divisor.
3. Cancel (if possible).
4. Multiply.

### Practice Exercise No. 12

Work out the following.

(1) $\frac{1}{3} \div 4$      (2) $\frac{1}{4} \div 6$      (3) $\frac{4}{5} \div 4$      (4) $\frac{3}{8} \div 9$

(5) $\frac{8}{9} \div 16$      (6) $\frac{18}{19} \div 9$      (7) $\frac{4}{5} \div 12$      (8) $\frac{6}{7} \div 27$

(9) $\frac{1}{5} \div 15$      (10) $\frac{3}{8} \div 8$      (11) $2\frac{1}{4} \div 3\frac{1}{2}$      (12) $3\frac{1}{8} \div 1\frac{1}{2}$

(13) $1\frac{1}{3} \div 2\frac{2}{3}$      (14) $\frac{1}{4} \div 1\frac{1}{4}$      (15) $8\frac{1}{2} \div 4\frac{1}{4}$      (16) $1\frac{1}{32} \div \frac{11}{16}$

(17) $4\frac{1}{3} \div \frac{2}{3}$      (18) $1\frac{1}{10} \div \frac{11}{15}$      (19) $1\frac{9}{10} \div 1\frac{3}{5}$      (20) $1\frac{3}{4} \div \frac{21}{32}$

(21) $2\frac{1}{3} \div \frac{2}{7}$      (22) $1\frac{6}{7} \div 3$      (23) $\frac{7}{8} \div 1\frac{3}{4}$      (24) $4\frac{1}{8} \div 5\frac{1}{2}$

(25) $1\frac{3}{5} \div \frac{2}{15}$

## 1.5 PRACTICAL APPLICATIONS EXERCISE

This is your first **Practical Applications Exercise**. These problems will give you an opportunity to exercise the math skills that you have just learned, problems similar to those you will find in your trade. You should attempt to work all of the problems.

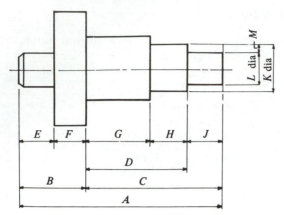

**Fig. 1–1**

Since many U.S. industries are now shifting to the metric system, there are problems included in metric units and in English, or Customary, units. Those problems in metric units have been indicated with an asterisk (*).

**(1)**   The component shown in Fig. 1–1 is made in several sizes, S1, S2, S3, etc. Neatly copy Table 1–2 and, from the information given, calculate the missing dimensions, thus completing the table. All dimensions are in inches (diagram not to scale).

**(2)**   Bars of material have to be cut to certain lengths using a circular saw, as shown in Fig. 1–2. Some material is wasted due to removal by the saw.

> **Question**   By how much will a bar be shortened if 5 components, each of length $\dfrac{3''}{4}$, are cut from it using an $\dfrac{1''}{8}$ wide saw?
>
> **Answer**   Material required for each component $=\dfrac{3''}{4} + \dfrac{1''}{8}$ (for saw cut)
>
> $$= \dfrac{6''}{8} + \dfrac{1''}{8} = \dfrac{7''}{8}$$
>
> Total length used $= 5 \times \dfrac{7''}{8} = \dfrac{35''}{8} = 4\dfrac{3''}{8}$

Find the length of material used for the following.

**(a)**   4 components $\dfrac{3''}{4}$ long using an $\dfrac{1''}{8}$ wide saw

**(b)**   9 components $\dfrac{7''}{16}$ long using a $\dfrac{1''}{16}$ wide saw

**TABLE 1–2**

| Size | A | B | C | D | E | F | G | H | J | K | L | M |
|------|---|---|---|---|---|---|---|---|---|---|---|---|
| S1 | 5 | $1\frac{1}{2}$ | — | $1\frac{3}{4}$ | $1\frac{1}{4}$ | — | $\frac{7}{8}$ | — | — | $1\frac{1}{4}$ | $\frac{3}{4}$ | — |
| S2 | $7\frac{7}{8}$ | — | $5\frac{1}{2}$ | — | — | $\frac{7}{16}$ | — | $1\frac{7}{16}$ | $1\frac{3}{16}$ | $1\frac{7}{8}$ | $\frac{11}{16}$ | — |
| S3 | — | $3\frac{7}{16}$ | $5\frac{1}{2}$ | — | $1\frac{3}{4}$ | — | $2\frac{7}{8}$ | $\frac{15}{16}$ | — | $3\frac{3}{4}$ | — | $\frac{7}{8}$ |
| S4 | $4\frac{1}{2}$ | — | $3\frac{3}{8}$ | — | $\frac{3}{16}$ | — | — | $1\frac{3}{8}$ | $\frac{5}{16}$ | — | $2\frac{7}{8}$ | $\frac{3}{16}$ |
| S5 | — | $2\frac{3}{32}$ | $4\frac{7}{8}$ | $3\frac{5}{32}$ | — | $1\frac{5}{16}$ | — | $2\frac{1}{8}$ | — | $1\frac{7}{16}$ | — | $\frac{3}{32}$ |
| S6 | $8\frac{1}{8}$ | $2\frac{1}{16}$ | — | $4\frac{7}{32}$ | $\frac{13}{32}$ | — | $2\frac{15}{16}$ | — | — | $3\frac{1}{16}$ | $2\frac{1}{4}$ | — |
| S7 | $9\frac{1}{2}$ | — | $7\frac{7}{16}$ | — | — | $\frac{15}{16}$ | — | $4\frac{1}{2}$ | $1\frac{3}{8}$ | $3\frac{3}{4}$ | $2\frac{15}{16}$ | — |
| S8 | $\frac{15}{16}$ | $\frac{3}{8}$ | — | $\frac{13}{32}$ | — | $\frac{3}{64}$ | — | $\frac{1}{8}$ | — | $\frac{7}{8}$ | $\frac{1}{2}$ | — |

**(c)** 7 components $\frac{5''}{8}$ long using a $\frac{1''}{16}$ wide saw

**(d)** 5 components $1\frac{1''}{8}$ long using a $\frac{1''}{16}$ wide saw

**(e)** 11 components $2\frac{1''}{4}$ long using an $\frac{1''}{8}$ wide saw

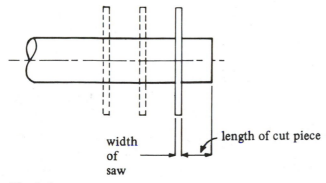

width of saw    length of cut piece

**Fig. 1–2**

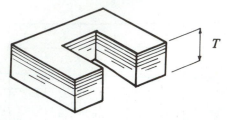

**Fig. 1–3**

(f)  8 components $1\frac{7''}{8}$ long using a $\frac{3''}{16}$ wide saw

(g)  3 components $4\frac{7''}{16}$ long using an $\frac{1''}{8}$ wide saw.

(3)  Answer the following questions using Fig. 1–3.

(a)  A block is to be made by fixing 44 plates, each $\frac{5''}{16}$ thick, together. What is the total thickness of the block $T$?

(b)  If the plates were $\frac{1''}{4}$ thick, what would $T$ be?

(c)  If another block (assembled) was measured and $T$ found to be $2\frac{5''}{8}$, find the thickness of the plates if 7 plates were used.

(d)  If $T = 3\frac{1''}{4}$ and the plates were $\frac{1''}{4}$ thick, how many plates were used?

(4)  Four balls $\frac{5''}{16}$ *radius* are placed together, as shown in Fig. 1–4. Calculate the distances $A$, $B$, $C$, and $D$.

(5)  Calculate $R$ and $L$ in Fig. 1–5.

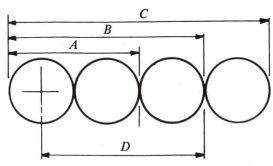

**Fig. 1–4**

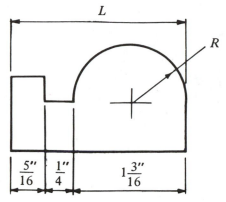

**Fig. 1–5**

**(6)** Calculate the dimensions $W$, $X$, $Y$, and $Z$ in Fig. 1–6.

**(7)** Calculate radius, $R$, and $H$ in Fig. 1–7.

**(8)** A worker is required to produce a pin $\dfrac{23''}{64}$ in diameter, but cannot find

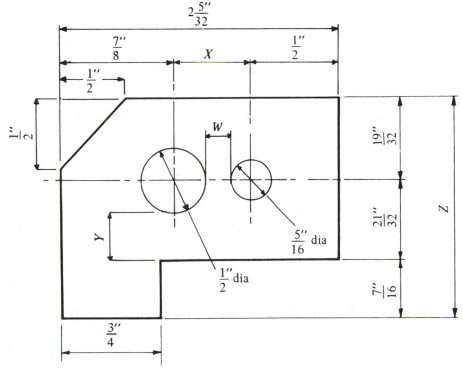

**Fig. 1–6**

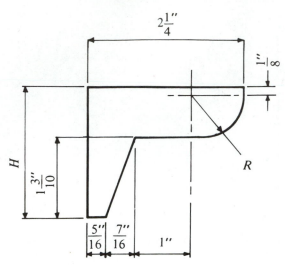

**Fig. 1–7**

one this size. The bars shown in Fig. 1–8 are the only ones available. A bar is selected and machined to the correct diameter. Which bar was chosen in order to waste the least amount of material?

**(9)**   Similar to Question 8. If we want a bar $\dfrac{17''}{32}$ in diameter, which of the following sizes would we choose to waste the least material?

$\dfrac{1''}{2}$ dia      $\dfrac{17''}{64}$ dia      $\dfrac{35''}{64}$ dia      $\dfrac{19''}{32}$ dia      $\dfrac{9''}{16}$ dia

**(10)**   A piece of metal $\dfrac{3''}{8}$ thick is placed between two blocks each $\dfrac{9''}{64}$ thick. What is the total thickness $T$ in Fig. 1–9?

**(11)**   A screw has 6 threads in $\dfrac{1''}{4}$. How many threads per inch?

**\*(12)**   Material for a component costs \$2.00 per meter; how much would $7\dfrac{1}{2}$ m cost?

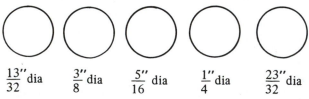

$\dfrac{13''}{32}$ dia      $\dfrac{3''}{8}$ dia      $\dfrac{5''}{16}$ dia      $\dfrac{1''}{4}$ dia      $\dfrac{23''}{32}$ dia

**Fig. 1–8**

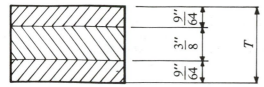

Fig. 1–9

**(13)** Find the cost of 350 chocolate bars at $12\frac{1}{2}¢$ each.

**(14)** Calculate $T$ in Fig. 1–10.

**(15)** A company produces 100 articles a day. If one worker produces 5, what fraction of the total output is this? Another worker produces 25 per day. What fraction of the total output is this?

**(16)** A saucepan has a small hole $\frac{9''}{16}$ in diameter in its bottom, as shown in Fig. 1–11. You have to repair it by placing a circular patch over the hole so that the patch extends $\frac{5''}{64}$ all around. What diameter will the patch have to be?

**(17)** Part of an electrical circuit is shown in Fig. 1–12.

Let $R$ be the combined resistance of $R_1$, $R_2$, and $R_3$, that is, the value of the single resistor that could exactly replace the three. $R$ can be obtained from the equation

$$\frac{1}{R} = \frac{1}{R_1} + \frac{1}{R_2} + \frac{1}{R_3}$$

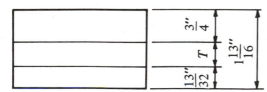

Fig. 1–10

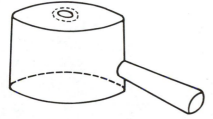

Fig. 1–11

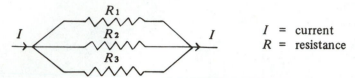

**Fig. 1–12**

$I$ = current
$R$ = resistance

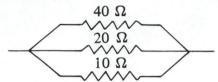

**Fig. 1–13**

*Example* Find the combined resistance of the resistors in Fig. 1–13.

*Answer* Let $R$ be the combined resistance,

$$\frac{1}{R} = \frac{1}{R_1} + \frac{1}{R_2} + \frac{1}{R_3}$$

$$\frac{1}{R} = \frac{1}{40} + \frac{1}{20} + \frac{1}{10} = \frac{1}{40} + \frac{2}{40} + \frac{4}{40} = \frac{7}{40}$$

Inverting (turning upside down) both sides,

$$\frac{R}{1} = \frac{40}{7}, \qquad R = 5\frac{5}{7}\,\Omega$$

Work out the combined resistance of the following.

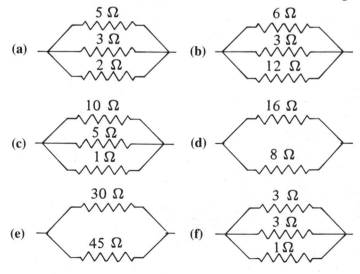

**(18)** **(a)** How many $\frac{1}{4}$ lb bags of nails could be obtained from $10\frac{1}{4}$ lb of nails?

**(b)** How many $\frac{1}{4}$ lb bags of seed could be obtained from $21\frac{1}{2}$ lb of seed?

*(c)** How many $\frac{1}{2}$ kg bags of sugar could be obtained from $35\frac{1}{2}$ kg of sugar?

## 1.6   MISCELLANEOUS EXERCISES

**(1)** Find the value of $5\frac{3}{8} + 2\frac{1}{16} - 1\frac{1}{4}$.

**(2)** Find the value of $14\frac{2}{3} \times 4\frac{1}{2} \div 1\frac{1}{2}$.

**(3)** Find the value of $3\frac{3}{8} + 2\frac{1}{16} - 5\frac{1}{8}$.

**(4)** Find the value of $12\frac{1}{2} \times 2\frac{7}{8} \div 1\frac{9}{16}$.

**(5)** Determine the value of $4\frac{1}{4} + 2\frac{3}{16} - 3\frac{1}{8}$.

**(6)** Determine the value of $3\frac{3}{4} \times 6\frac{2}{3} \div 1\frac{1}{4}$.

**(7)** Determine the value of $1\frac{7}{15} \times 2\frac{5}{11} - 4\frac{4}{7} \div 2\frac{2}{3}$.

**(8)** Determine the value of $\dfrac{1}{\dfrac{3}{4} + \dfrac{6}{7}} - \dfrac{2\frac{1}{3} - 1\frac{3}{5}}{6\frac{3}{5}}$.

**(9)** Simplify and evaluate $1\frac{7}{16} \div 1\frac{1}{2} - \frac{3}{4}$.

**(10)** **(a)** Simplify $\left(1\frac{1}{2} + \frac{1}{6}\right) \times 2\frac{2}{5}$.

**(b)** Simplify $\frac{7}{12} \div \frac{3}{4} - \frac{7}{18}$.

**(11)** Simplify $\left(\frac{2}{3} + \frac{2}{5}\right) \times \frac{5}{8}$.

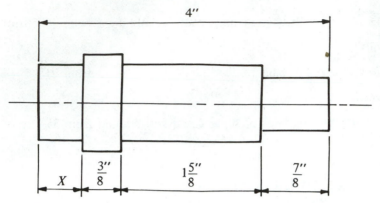

**Fig. 1–14**

**(12)** You are required to machine dimension $X$ to length. Work out what length $X$ will be in Fig. 1–14.

**(13)** Figure 1–15 shows a shaft. What is length $X$?

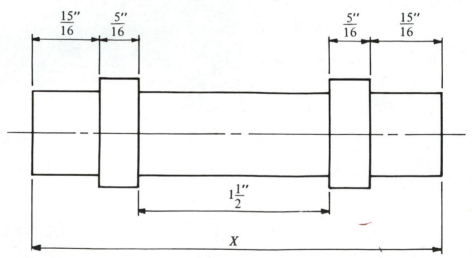

**Fig. 1–15**

# 2

# Decimal Fractions

---

**Self-Assessment Test No. 1**

Change the following common fractions to decimal fractions.

(1)  $3\frac{1}{10}$    (2)  $5\frac{25}{100}$    (3)  $16\frac{2}{100}$

(4)  $7\frac{7}{1000}$    (5)  $17\frac{33}{100}$    (6)  $8\frac{22}{1000}$

Change the following decimal fractions to common fractions.

(7)  1.3    (8)  1.8    (9)  3.875

(10)  10.75    (11)  8.45    (12)  11.12

## 2.1  INTRODUCTION TO DECIMAL FRACTIONS[1]

Consider the number 3413.

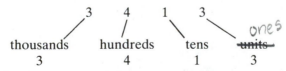

$$
\begin{array}{cccc}
3 & 4 & 1 & 3 \\
\text{thousands} & \text{hundreds} & \text{tens} & \text{ones} \\
3 & 4 & 1 & 3
\end{array}
$$

The use of a decimal point allows us to write tenths, hundredths, thousandths, etc., in a number. In decimal numbers, all fractions are written to the *right* of the point.

---

[1] The first reference to the *optional* use of a calculator is made in Section 2.5 of this chapter.

Let's look at a decimal number.

| 1 | 3 | 2 | 9 | 8 | 2 |
|---|---|---|---|---|---|
| hundreds | tens | units | tenths | hundredths | thousandths |
| (100) | (10) | (1) | $\left(\dfrac{1}{10}\right)$ | $\left(\dfrac{1}{100}\right)$ | $\left(\dfrac{1}{1000}\right)$ |
| 1 | 3 | 2 | 9 | 8 | 2 |

$$0.9 \text{ means } \frac{9}{10}$$

$$0.98 \text{ means } \frac{98}{100}$$

$$0.982 \text{ means } \frac{982}{1000}$$

Consider the common fraction $13\frac{3}{10}$; this is the same as the decimal number 13.3.

$$13.37 = 13\frac{37}{100} \qquad 16\frac{451}{1000} = 16.451$$

$$2\frac{1}{100} = 2.01 \qquad 3\frac{9}{1000} = 3.009$$

## Practice Exercise No. 1

Express the following as decimal numbers.

(1)  $1\frac{1}{10}$  (2)  $4\frac{3}{10}$  (3)  $5\frac{51}{100}$  (4)  $2\frac{71}{100}$

(5)  $3\frac{911}{1000}$  (6)  $2\frac{9}{10}$  (7)  $3\frac{37}{100}$  (8)  $3\frac{7}{100}$

(9)  $3\frac{7}{1000}$  (10)  $2\frac{49}{1000}$  (11)  $3\frac{85}{1000}$  (12)  $7\frac{7}{10}$

(13)  $2\frac{33}{100}$  (14)  $1\frac{1}{1000}$  (15)  $12\frac{19}{100}$

Now consider the number 13.03.

This means $13 + 0 \times \frac{1}{10}$'s $+ 3 \times \frac{1}{100}$'s or $13 + \frac{03}{100}$.

$$\therefore 13.03 = 13\frac{3}{100}$$

Consider the number 13.009.

This means $13 + 0 \times \frac{1}{10}$'s $+ 0 \times \frac{1}{100}$'s $+ 9 \times \frac{1}{1000}$'s or $13 + \frac{009}{1000}$.

$$\therefore 13.009 = 13\frac{9}{1000}$$

Similarly, $13.039 = 13\frac{39}{1000}$.

## Practice Exercise No. 2

Express the following decimal numbers as common fractions.

| | | | | |
|---|---|---|---|---|
| **(1)** 1.1 | **(2)** 1.11 | **(3)** 1.111 | **(4)** 1.03 | **(5)** 1.003 |
| **(6)** 1.714 | **(7)** 1.3 | **(8)** 2.93 | **(9)** 4.4 | **(10)** 7.771 |
| **(11)** 1.93 | **(12)** 2.375 | **(13)** 6.5 | **(14)** 1.25 | **(15)** 1.030 |
| **(16)** 2.41 | **(17)** 1.937 | **(18)** 1.19 | **(19)** 16.75 | **(20)** 19.125 |

Adding the figure 0 to the end of a decimal number does not alter the value of the number. For example,

$$25.1 \text{ means } 25 + 1 \times \frac{1}{10}$$

$$25.10 \text{ means } 25 + 1 \times \frac{1}{10} + 0 \times \frac{1}{100}$$

$$25.100 \text{ means } 25 + 1 \times \frac{1}{10} + 0 \times \frac{1}{100} + 0 \times \frac{1}{1000}$$

$$\therefore 25.1 = 25.10 = 25.100$$

## 2.2 SIMPLIFYING OPERATIONS FOR DECIMALS

### Decimal Places

A number with 2 figures to the *right* of the decimal point has 2 decimal places. Thus,

4.91 has 2 decimal places

4.9136 has 4 decimal places

4.1 has 1 decimal place

## Multiplication of Numbers by Powers of 10 (10, 100, 1000, etc.)

When multiplying a number by a power of 10, move the decimal point as many places to the *right* as there are 0's in the multiplier. If there are not enough figures, then 0's must be added.

$$0.35 \times 10 = .3\,5 = 3.5$$
$$0.0637 \times 100 = .0\,6\,37 = 6.37$$
$$6.3 \times 1000 = 6.\,3\,0\,0 = 6300$$

### Practice Exercise No. 3

Work out the following.

| | | |
|---|---|---|
| **(1)** 0.7 × 10 | **(2)** 0.7 × 100 | **(3)** 0.7 × 1000 |
| **(4)** 0.313 × 10 | **(5)** 0.4134 × 1000 | **(6)** 0.033 × 100 |
| **(7)** 0.612 × 1000 | **(8)** 1.477 × 10 | **(9)** 47.63 × 10 |
| **(10)** 88.23 × 1000 | **(11)** 49.93 × 100 | **(12)** 18.776 × 1000 |
| **(13)** 0.003 × 100 | **(14)** 121.9 × 100 | **(15)** 1.775 × 10 |
| **(16)** 0.004 × 100 | **(17)** 0.123 × 10 000 | **(18)** 1.995 × 10 |
| **(19)** 6.667 × 100 | **(20)** 4.9 × 10 000 | |

## Division of Numbers by Powers of 10 (10, 100, 1000, etc.)

When dividing by numbers like 10, 100, 1000, etc., move the decimal point as many places to the *left* as there are 0's in the divisor. If there are not enough figures, then 0's must be placed in front of the number to fill up the empty spaces. For example,

$$\frac{7}{10} = 7 = 0.7 \qquad \frac{45}{10} = 4\,5 = 4.5$$

$$\frac{15}{100} = 1\,5 = 0.15 \qquad \frac{15}{10\,000} = 0\,0\,1\,5 = 0.0015$$

$$\frac{3}{1000} = 0\,0\,3 = 0.003 \qquad \frac{4}{100} = 0\,4 = 0.04$$

## Practice Exercise No. 4

Work out the following (give your answer in decimal form).

(1) $\dfrac{9}{10}$ 　　(2) $\dfrac{9}{100}$ 　　(3) $\dfrac{442}{1000}$ 　　(4) $11\dfrac{9}{100}$

(5) $14\dfrac{7}{10\ 000}$ 　　(6) $19\dfrac{3}{100}$ 　　(7) $\dfrac{900}{10\ 000}$ 　　(8) $\dfrac{175}{1000}$

(9) $\dfrac{35}{10}$ 　　(10) $\dfrac{9946}{100}$ 　　(11) $\dfrac{4.75}{100}$ 　　(12) $\dfrac{0.032}{10}$

(13) $11\dfrac{1}{1000}$ 　　(14) $2\dfrac{3}{100}$ 　　(15) $\dfrac{1.1}{1000}$ 　　(16) $\dfrac{9.375}{100}$

(17) $\dfrac{4.421}{10}$ 　　(18) $\dfrac{19.773}{1000}$ 　　(19) $4\dfrac{75}{1000}$ 　　(20) $\dfrac{16}{100}$

## Conversion of Decimal Fractions to Fractions or Mixed Numbers

Here are some examples.

$$0.65 = \frac{\cancel{65}^{13}}{\cancel{100}_{20}} = \frac{13}{20} \qquad 1.13 = 1\frac{13}{100}$$

$$0.25 = \frac{\cancel{25}^{1}}{\cancel{100}_{4}} = \frac{1}{4} \qquad 2.15 = 2\frac{\cancel{15}^{3}}{\cancel{100}_{20}} = 2\frac{3}{20}$$

$$0.2 = \frac{\cancel{2}^{1}}{\cancel{10}_{5}} = \frac{1}{5} \qquad 3.75 = 3\frac{\cancel{75}^{3}}{\cancel{100}_{4}} = 3\frac{3}{4}$$

To convert a decimal fraction to a common fraction, rewrite the decimal fraction as a fraction with the denominator as a power of 10 (i.e., 10, 100, 1000, etc.) and then cancel to the lowest form.

## Practice Exercise No. 5

Convert the following decimal fractions to common fractions or mixed numbers.

(1) 0.85 　　(2) 0.6 　　(3) 0.5 　　(4) 0.75 　　(5) 0.99

(6) 0.977 　　(7) 0.875 　　(8) 0.8 　　(9) 0.9 　　(10) 3.25

(11) 35.46 　　(12) 16.125 　　(13) 0.12 　　(14) 0.375 　　(15) 0.031 25

(16) 1.3 　　(17) 10.45 　　(18) 19.1 　　(19) 4.39 　　(20) 10.625

*Self-Assessment Test No. 2*

(1)   Find the sum of 1.013, 13, and 0.013.

(2)   Find the sum of 1000, 1.01, and 3.125.

(3)   Find the sum of 0.13, 0.005, and 16.95.

(4)   Find the sum of 146, 27, and 99.13.

(5)   Subtract 1.01 from 2.

(6)   Subtract 99.57 from 200.

(7)   Subtract 0.012 from 1.

(8)   Subtract 1.137 from 14.

(9)   Subtract 10.34 from 11.01.

(10)   Subtract 0.075 from 12.

## 2.3   ADDITION AND SUBTRACTION OF DECIMALS

To add 14.576 and 13.334, write down the numbers with the decimal points *directly in line* and then add the numbers in the normal way.

$$\begin{array}{r} 14.576 \\ + \ 13.334 \\ \hline 27.910 \end{array}$$

The decimal point in the answer is placed directly in line with the decimal point of the numbers being added.

*Example*   Add 18.3 and 14.796.

$$\begin{array}{r} 18.3 \\ + \ 14.796 \\ \hline \phantom{00000} \end{array}$$

**Hint:**   *Adding 0's to the right-hand side of a decimal number does not alter the value of that number; 18.3 could be written as 18.300 to make the addition easier.*

$$\begin{array}{r} 18.300 \\ + \ 14.796 \\ \hline 33.096 \end{array}$$

## Practice Exercise No. 6

Work out the following.

| (1) | 11.013 | (2) | 210.13 | (3) | 444.0 | (4) | 79.66 |
|---|---|---|---|---|---|---|---|
| | 9.732 | | 10.912 | | 6.79 | | 0.093 |
| + | 6.441 | + | 6.09 | + | 41.32 | + | 101.32 |

| (5) | 662.31 | (6) | 644.3 | (7) | 61.32 | (8) | 14.733 |
|---|---|---|---|---|---|---|---|
| | 0.404 | | 2.613 | | 0.104 | | 0.091 |
| + | 7921.61 | + | 0.075 | + | 64.9 | + | 9.76 |

| (9) | 4.216 | (10) | 99.47 | (11) | 133.7 | (12) | 0.036 |
|---|---|---|---|---|---|---|---|
| | 273.32 | | 0.062 | | 4.601 | | 4.9 |
| + | 0.0916 | + | 7.93 | + | 51.97 | + | 512.33 |

(13) Find the sum of 3.02, 45.61, 17, and 83.33.

(14) Find the sum of 3.775, 16.251, and 95.43.

(15) Find the sum of 14.664, 7, 16.559, and 143.

(16) Find the sum of 0.009, 76, 0.0321, 66.109, and 0.110.

(17) Find the sum of 111.03, 72, 61, 0.009, and 16.45.

(18) Find the sum of 77.93, 16, 0.009, and 101.

(19) Find the sum of 1003.9, 2.88, and 37.775.

(20) Find the sum of 16.36, 14.003, and 142.

To subtract one decimal number from another, both decimal numbers must have the decimal points directly in line (as in addition). For example,

$$
\begin{array}{r}
303.43 \\
- \quad 20.88 \\
\hline
282.55
\end{array}
$$

## Practice Exercise No. 7

Work out the following.

| (1) | 144.8 | (2) | 19.297 | (3) | 14.912 | (4) | 19.732 |
|---|---|---|---|---|---|---|---|
| − | 72.733 | − | 8.613 | − | 9.399 | − | 9.813 |

| (5) | 88.366 | (6) | 0.319 | (7) | 114.93 | (8) | 19.113 |
|-----|--------|-----|-------|-----|--------|-----|--------|
| | − 63.77 | | − 0.29 | | − 73.89 | | − 8.889 |

| (9) | 14.9 | (10) | 77.942 | (11) | 609.3 | (12) | 16.39 |
|-----|------|------|--------|------|-------|------|-------|
| | − 0.099 | | − 66.99 | | − 98.95 | | − 4.97 |

(13) Subtract 1.304 from 99.

(14) Subtract 1.034 from 2.

(15) Subtract 62.43 from 101.854.

(16) Subtract 10.89 from 11.03.

(17) Subtract 14.997 from 64.

(18) Subtract 0.995 from 1.003.

(19) Subtract 17.332 from 19.5.

(20) Subtract 77.98 from 810.

# 2.4 PRACTICAL APPLICATIONS
## EXERCISE NO. 1

*(1) Normal body temperature is 37°C. A person running a temperature of 38.3°C is how many degrees above normal?

(2) A table has dimensions 2.95 yd × 1.45 yd. What is the perimeter of the table?

(3) You are planning to make a fruit bowl as shown in Fig. 2–1. Calculate
(a) Internal diameter $a$.
(b) Internal radius $r$.
(c) Internal depth $d$.

*(4) A component, as shown in Fig. 2–2, is to be produced on a lathe. How much must the cutting tools $A$ and $B$ be wound (or fed) to produce the sizes stated?

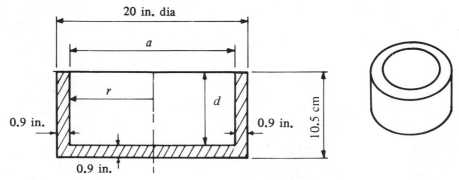

Fig. 2–1

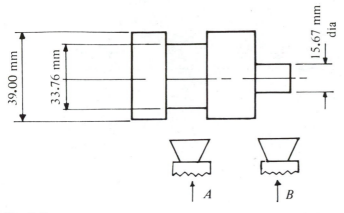

**Fig. 2–2**

*(5) A component is to be produced as shown in Fig. 2–3. Calculate
(a) $X$ and $Y$.
(b) $Z$ (the hole is equidistant from faces $A$ and $B$).

(6) A drawing gives the length of a bar as 55.78″ ± 0.05″.
(a) What is the length of the largest bar that is acceptable?
(b) What is the length of the shortest bar that is acceptable?

(7) A component consists of a flat plate with two pins positioned and fixed as shown in Figs. 2–4 and 2–5. From the dimensions given, calculate
(a) $A$, $B$, $C$, $D$, $E$, and $F$.
(b) The radii of the two pins.
(c) The perimeter of the flat plate.

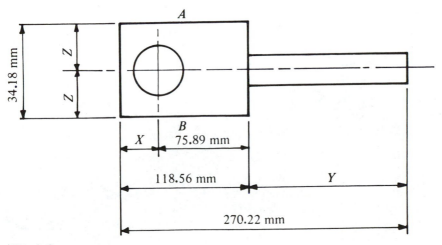

**Fig. 2–3**

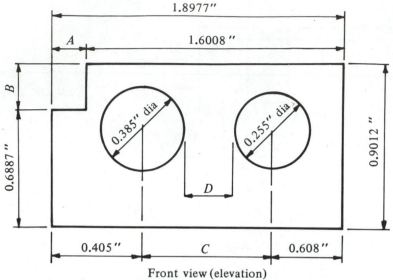

Front view (elevation)

**Fig. 2–4**

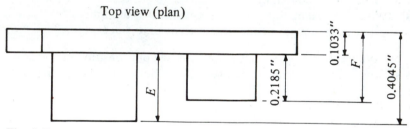

Top view (plan)

**Fig. 2–5**

*(8)  What is the inside diameter of a cylinder that has an external diameter of 1.34 m and is made of material 0.85 cm thick?
(Note the units!)

---

*Self-Assessment Test No. 3*

Work out the following.

  **(1)**  1.97 × 3.14        **(2)**  19.35 × 1.03        **(3)**  17.96 × 14.32

  **(4)**  1.003 × 10.5       **(5)**  16.9 × 1.250       **(6)**  112 × 0.013

  **(7)**  The cost of 7.6 lb of material at $1.23 per lb.

  **(8)**  The cost of 95.8 lb of material at $0.23½ per lb.

 **\*(9)**  Convert 20.5 in. to millimeters (1 in. = 25.4 mm).

**\*(10)**  Eleven rods, each 44.75 mm long, are joined together. What is the total length of these rods?

## 2.5 MULTIPLICATION OF DECIMALS

**Calculators**

If your instructor permits the use of calculators, this may be a good point to turn to Appendix C and begin to learn how to use your calculator effectively.

Section C.1 is all that you will need to look at first. At the appropriate places throughout the rest of the text, you will be referred to other sections of Appendix C for *Calculator Hints*. The *Hint* will be of direct use in that chapter and section, and will probably make better sense when you begin using the Appendix.

When multiplication involves decimals, start by ignoring the decimal point and proceed as in multiplication of whole numbers. When an answer has been obtained, count the total number of decimal places in the numbers being multiplied. There must be the same number of decimal places in the answer.

*Example (1)*   $21.4 \times 3.2$

$$
\begin{array}{r}
21.4 \\
\times\ 3.2 \\
\hline
428 \\
6420 \\
\hline
68.48 \\
\hline
\end{array}
$$

Total number of decimal places in question = 2

Total number of decimal places in answer = 2

*Answer*   $21.4 \times 3.2 = 68.48$

*Example (2)*   $2.75 \times 12$

$$
\begin{array}{r}
2.75 \\
\times\ \ 12 \\
\hline
550 \\
2750 \\
\hline
33.00 \\
\hline
\end{array}
$$

Total number of decimal places in question is $2 + 0 = 2$.

Total number of decimal places in answer = 2.

*Answer*   $2.75 \times 12 = 33.00$ or 33

*Example (3)*   $30.12 \times 21.4$

$$
\begin{array}{r}
30.12 \\
\times\ 21.4 \\
\hline
12048 \\
30120 \\
602400 \\
\hline
644.568 \\
\hline
\end{array}
$$

Total number of decimal places in question = 3

Total number of decimal places in answer = 3

*Answer* 30.12 × 21.4 = 644.568

*Example (4)* 0.003 × 0.004

$$
\begin{array}{r}
0.003 \\
\times\,0.004 \\
\hline
.\overbrace{000012} \\
\hline
\end{array}
$$

In this case there are, initially, only 2 figures in the answer but six decimal places in the question. Zeros have to be inserted to place the decimal point in the correct position.

*Answer* 0.003 × 0.004 = 0.000 012

## Practice Exercise No. 8

Work out the following.

**(1)** 3.7 × 0.21     **(2)** 6.5 × 7.3     **(3)** 8.9 × 9.1

**(4)** 0.02 × 760     **(5)** 77.7 × 16.3     **(6)** 0.0035 × 0.0021

**(7)** 94.3 × 6.72     **(8)** 49.77 × 0.0312     **(9)** 16.993 × 0.004

**(10)** 18.912 × 0.003     **(11)** 7.663 × 1.95     **(12)** 14.7 × 16.31

**(13)** 0.0337 × 0.0135     **(14)** 0.0375 × 1.22     **(15)** 1.776 × 0.023

**(16)** 121 × 3.13

## 2.6 PRACTICAL APPLICATIONS EXERCISE NO. 2

*(1) A person orders a set of rectangular sheets of material that have varying widths and lengths. (See Fig. 2–6.)

Given that the area of a rectangle, $A$, is equal to length times perpendicular height, or $= L \times W$.

(a) Neatly copy the following table.

(b) Calculate the area in square meters for each of the sheets listed in the table.

(c) If the cost of material is $2.74 per square meter, calculate the cost of each of these sheets.

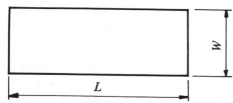

Fig. 2–6

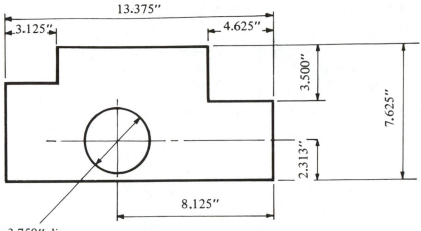

**Fig. 2–7**

$$\text{Cost/Sheet} = \text{Area (m}^2) \times \text{Cost/m}^2 \text{ (\$)}$$

| Length (m) | Width (m) | Area (m²) | Cost ($) |
|:---:|:---:|:---:|:---:|
| .32 | 0.75 | | |
| 1.78 | 1.39 | | |
| 2.45 | 1.98 | | |
| 2.92 | 2.07 | | |
| 3.06 | 2.78 | | |

*(2)  A component is designed as shown in Fig. 2–7.

However, all the machinery and tools to produce this part are calibrated in metric units. Sketch the component, giving dimensions in millimeters. Give dimensions correct to 2 decimal places (1″ = 25.40 mm), by multiplying each dimension by 25.40 mm.

*(3)  Sketch the component in Fig. 2–8 giving dimensions in inches. Give values correct to 3 decimal places (1 mm = 0.0394″).

---

*Self-Assessment Test No. 4*

Work out the following giving the answer to 3 decimal places where appropriate.

**(1)**  16.34 ÷ 0.7     **(2)**  8.13 ÷ 0.03     **(3)**  18.77 ÷ 1.5

**(4)**  19.18 ÷ 2.8     **(5)**  77.65 ÷ 7.6     **(6)**  19.3 ÷ 0.99

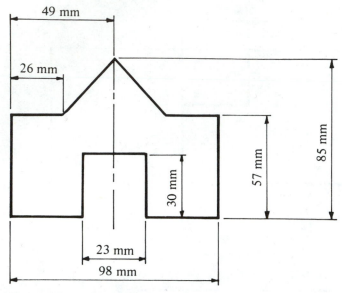

**Fig. 2–8**

## 2.7 DIVISION OF DECIMALS

### Long Division (Review)

*Example (1)*   928 ÷ 16

$$
\begin{array}{r}
58 \\
16\overline{)928} \\
80\downarrow \\
\hline
128 \\
128 \\
\hline
000
\end{array}
$$

*Answer*   58

*Example (2)*   14 592 ÷ 32

$$
\begin{array}{r}
456 \\
32\overline{)14592} \\
128\downarrow\downarrow \\
\hline
179 \\
160\downarrow \\
\hline
192 \\
192 \\
\hline
000
\end{array}
$$

*Answer*   456

*Example (3)*   45 435 ÷ 15

$$
\begin{array}{r}
3029 \\
15\overline{)45435} \\
45\phantom{000} \\
\overline{\phantom{0}0043} \\
30\phantom{0} \\
\overline{\phantom{00}135} \\
135 \\
\overline{\phantom{0}000}
\end{array}
$$

*Answer*   3029

## Practice Exercise No. 9

Work out the following by long division.

|     |              |      |               |      |              |
|-----|--------------|------|---------------|------|--------------|
| **(1)** | 7344 ÷ 16 | **(2)** | 8025 ÷ 25 | **(3)** | 39 797 ÷ 17 |
| **(4)** | 13 468 ÷ 13 | **(5)** | 8910 ÷ 45 | **(6)** | 75 537 ÷ 21 |
| **(7)** | 25 740 ÷ 33 | **(8)** | 12 482 ÷ 79 | **(9)** | 24 738 ÷ 31 |
| **(10)** | 14 084 ÷ 14 | **(11)** | 78 383 ÷ 761 | **(12)** | 10 387 ÷ 13 |

## Dividing a Decimal Number by a Whole Number

*Example (1)*   2.17 ÷ 7

Write down the problem in the normal way and place a decimal point on the answer line directly above the decimal point of the number being divided. That is,

$$
7\overline{)2.17}
$$

Then divide as before.

$$
\begin{array}{r}
0.31 \\
7\overline{)2.17} \\
2\ 1\phantom{0} \\
\overline{0\ 07} \\
7 \\
\overline{0}
\end{array}
$$

*Answer*   0.31

A better method of deciding where to place the decimal point is to count the same number of decimal places in the answer as in the number below it—the 22.88 in Example (2)—after doing the long division as before.

In Example (2) below, the 22.88 has 2 decimal places, so the answer will also have 2 decimal places.

***Example (2)*** 22.88 ÷ 13

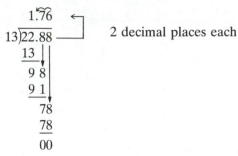

2 decimal places each

*Answer* 1.76

## Practice Exercise No. 10

Work out the following by long division.

| | | |
|---|---|---|
| **(1)** 19.8 ÷ 22 | **(2)** 75.43 ÷ 19 | **(3)** 142.56 ÷ 33 |
| **(4)** 198.03 ÷ 21 | **(5)** 12.114 ÷ 18 | **(6)** 1.08 ÷ 18 |
| **(7)** 53.44 ÷ 32 | **(8)** 45.653 ÷ 71 | **(9)** 51.66 ÷ 14 |
| **(10)** 117.64 ÷ 17 | **(11)** 48.64 ÷ 19 | **(12)** 221.12 ÷ 32 |

## Dividing a Decimal Number by a Decimal Number

***Example (1)*** $\dfrac{27.95}{1.3}$

When the divisor (dividing number) is not a whole number, it must be changed into one. To convert 1.3 to a whole number, we must multiply it by 10, that is, move the decimal point one place to the right. Therefore,

$$1.3 \times 10 = 1.3\!\!\!\nearrow = 13$$

If we do this, however, we must also multiply 27.95 by 10. Then

$$27.95 \times 10 = 27.9\!\!\!\nearrow 5 = 279.5$$

$$\therefore \frac{27.95}{1.3} = \frac{27.95 \times 10}{1.3 \times 10} = \frac{279.5}{13}$$

We can then divide as before.

$$
\begin{array}{r}
21.5 \\
13\overline{)279.5} \\
\underline{26} \\
19 \\
\underline{13} \\
6\ 5 \\
\underline{6\ 5} \\
0\ 0
\end{array}
$$

1 decimal place each

*Answer*   21.5

*Example (2)*   $0.6304 \div 0.32$

$$\frac{0.6304}{0.32} = \frac{0.6304 \times 100}{0.32 \times 100} = \frac{63.04}{32}$$

$$
\begin{array}{r}
1.97 \\
32\overline{)63.04} \\
\underline{32} \\
31\ 0 \\
\underline{28\ 8} \\
2\ 24 \\
\underline{2\ 24} \\
0\ 00
\end{array}
$$

2 decimal places each

*Answer*   1.97

## Practice Exercise No. 11

Work out the following.

| | | |
|---|---|---|
| **(1)**  $5.840 \div 1.6$ | **(2)**  $8.358 \div 2.1$ | **(3)**  $59.80 \div 6.5$ |
| **(4)**  $1.728 \div 0.18$ | **(5)**  $1.4592 \div 0.32$ | **(6)**  $6.194 \div 0.19$ |
| **(7)**  $7.588 \div 0.14$ | **(8)**  $18.328 \div 2.9$ | **(9)**  $0.8260 \div 3.5$ |
| **(10)**  $15.216 \div 1.6$ | **(11)**  $71.44 \div 0.76$ | **(12)**  $4.154 \div 0.62$ |

## Division with Remainders

In the previous problems, all the numbers have been divisible without there being a "remainder," which does not always happen.

***Example (1)***   $\dfrac{0.36}{0.8}$

$$\frac{0.36}{0.8} = \frac{0.3\!\overset{\frown}{6}}{0.\overset{\curvearrowleft}{8}} = \frac{3.6}{8}$$

$$8\overline{)3.6} \\ \underline{3\ 2} \\ 4$$
$.4$

When this stage has been reached, a zero must be added to the number being divided; that is,

$$8\overline{)3.60} \\ \underline{3\ 2} \\ 40 \\ \underline{40} \\ 00$$
$0.45$

2 decimal places each, re-member the decimal places are counted *last*

***Answer***   0.45

***Example (2)***   $\dfrac{0.3}{1.6}$

$$\frac{0.3}{1.6} = \frac{0.\overset{\frown}{3}}{1.\overset{\frown}{6}} = \frac{3}{16}$$

$$16\overline{)3.0000} \\ \underline{1\ 6} \\ 1\ 40 \\ \underline{1\ 28} \\ 120 \\ \underline{112} \\ 80 \\ \underline{80} \\ 00$$
$0.1875$

***Answer***   0.1875

## Rounding

Sometimes the division appears never to end and we must give our answer *correct to a certain number of decimal places.* For example,

$\dfrac{2}{3}$ means $2 \div 3$

$$3\overline{)2.000} \\ \underline{1\ 8} \\ 20 \\ \underline{18} \\ 20 \\ \underline{18} \\ 2$$
$0.666\ 666\ldots$

We may have to solve *"correct to three decimal places."*

If the figure *after* the last decimal place we are told to work to is *5 or more*, the value of the last decimal place is *increased by 1*.

If the figure *after* the decimal place required is *4 or less*, the value of the last decimal place *is not changed*.

$$\therefore \frac{2}{3} = 0.667 \text{ (to 3 decimal places)}$$

Here are some examples.

Write 21.3377 correct to 3 decimal places.   *Answer*   21.338

Write 13.5743 correct to 2 decimal places.   *Answer*   13.57

Write 14.8910 correct to 1 decimal place.   *Answer*   14.9

Write 99.139 correct to 2 decimal places.   *Answer*   99.14

Write 0.0293 correct to 2 decimal places.   *Answer*   0.03

## Practice Exercise No. 12

Write the following numbers correct to the number of decimal places (d.p.) stated.

**(1)**   31.9967 to   **(a)** 3 d.p.   **(b)** 1 d.p.   **(c)** 2 d.p.

**(2)**   77.4489 to   **(a)** 1 d.p.   **(b)** 2 d.p.   **(c)** 3 d.p.

**(3)**   0.03725 to   **(a)** 4 d.p.   **(b)** 2 d.p.   **(c)** 3 d.p.

**(4)**   101.7791 to   **(a)** 1 d.p.   **(b)** 2 d.p.   **(c)** 3 d.p.

## Division with Rounding

Sometimes an answer may appear to never end in a decimal division.

*Example (1)*   100 ÷ 33

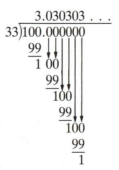

It may be necessary to correct to a certain number of decimal places. That is,

$$\frac{100}{33} = 3.030303 \ldots$$

$$= 3.030 \text{ (correct to 3 decimal places)}$$

Remember! Always work to one more place than the number of decimal places required and then round off the last figure accordingly.

*Example (2)*   13.7 ÷ 1.7 (Give answer to 3 decimal places if necessary.)

$$\frac{13.7}{1.7} = \frac{13.7}{1.7} = \frac{137}{17}$$

```
        8.0588 . . .
17)137.0000
   136
     1 00
       85
      150
      136
      140
      136
        4
```

*Answer*   8.059 (to 3 d.p.)

## Practice Exercise No. 13

Work out the following, giving the answer correct to 3 decimal places (if necessary).

| | | |
|---|---|---|
| **(1)** 21.4 ÷ 1.7 | **(2)** 323 ÷ 3.1 | **(3)** 18.91 ÷ 1.3 |
| **(4)** 16.72 ÷ 1.9 | **(5)** 321 ÷ 1.4 | **(6)** 79.6 ÷ 2.9 |
| **(7)** 18.62 ÷ 33 | **(8)** 41.7 ÷ 3.5 | **(9)** 167 ÷ 0.93 |
| **(10)** 1304 ÷ 1.7 | **(11)** 29.14 ÷ 0.31 | **(12)** 14.63 ÷ 1.5 |
| **(13)** 1.87 ÷ 39 | **(14)** 17.66 ÷ 1.1 | **(15)** 2094 ÷ 193 |
| **(16)** 1777 ÷ 34.2 | **(17)** 14.95 ÷ 10.3 | **(18)** 16.43 ÷ 9.9 |
| **(19)** 18.94 ÷ 7.21 | **(20)** 14.44 ÷ 99.9 | |

---

*Self-Assessment Test No. 5*

Convert the following common fractions to decimal fractions (give the answer to 3 decimal places).

| | | |
|---|---|---|
| **(1)** $\frac{1}{5}$ | **(2)** $\frac{7}{8}$ | **(3)** $\frac{11}{40}$ |
| **(4)** $\frac{5}{16}$ | **(5)** $\frac{3}{7}$ | **(6)** $\frac{9}{17}$ |

Convert the following decimal fractions to the nearest 16th of an inch.

**(7)**   8.753″                    **(8)**   9.871″

Convert the following decimal fractions to the nearest 64th of an inch.

**(9)**   17.5339″                    **(10)**   24.7935″

# 2.8 CONVERSIONS BETWEEN FRACTIONS AND DECIMALS

## Conversion of Common Fractions to Decimal Fractions

If a common fraction has to be converted to decimal form, you must divide the numerator (top number) by the denominator (bottom number).

*Example (1)*  $\frac{1}{2}$ means $1 \div 2$

$$
\begin{array}{r}
0.5 \\
2\overline{)1.0} \\
\underline{1.0} \\
00
\end{array}
\qquad \frac{1}{2} = 0.5
$$

*Example (2)*  $\frac{1}{4}$ means $1 \div 4$

$$
\begin{array}{r}
0.25 \\
4\overline{)1.00} \\
\underline{8\downarrow} \\
20 \\
\underline{20} \\
00
\end{array}
\qquad \frac{1}{4} = 0.25
$$

*Example (3)*  $\frac{3}{8}$ means $3 \div 8$

$$
\begin{array}{r}
0.375 \\
8\overline{)3.000} \\
\underline{24\downarrow} \\
60 \\
\underline{56\downarrow} \\
40 \\
\underline{40} \\
00
\end{array}
\qquad \frac{3}{8} = 0.375
$$

## Practice Exercise No. 14

Convert the following common fractions to decimal fractions.

(1) $\frac{1}{5}$     (2) $\frac{5}{8}$     (3) $\frac{7}{8}$     (4) $\frac{19}{50}$     (5) $\frac{1}{8}$

(6) $\frac{9}{20}$     (7) $\frac{4}{5}$     (8) $\frac{3}{4}$     (9) $\frac{3}{5}$     (10) $\frac{9}{40}$

(11) $\frac{3}{10}$     (12) $\frac{52}{100}$     (13) $\frac{93}{1000}$     (14) $\frac{516}{10\,000}$     (15) $\frac{79}{100}$

As before, it will sometimes be necessary to round the conversion (division) to a certain number of decimal places.

### Practice Exercise No. 15

Convert the following fractions to decimal fractions (give the answer to 3 decimal places).

(1) $\dfrac{1}{6}$    (2) $\dfrac{1}{3}$    (3) $\dfrac{5}{6}$    (4) $\dfrac{2}{7}$    (5) $\dfrac{4}{9}$    (6) $\dfrac{5}{12}$

(7) $\dfrac{1}{16}$    (8) $\dfrac{5}{16}$    (9) $\dfrac{11}{12}$    (10) $\dfrac{2}{11}$    (11) $\dfrac{7}{16}$    (12) $\dfrac{5}{7}$

## Conversion of Decimal Fractions to Common Fractions

### Exact

We saw an easy way to do this conversion in Section 2.1; that is,

$$2.78 = 2 + \frac{78}{100} = 2\frac{78}{100} = 2\frac{\cancel{78}^{39}}{\cancel{100}_{50}} = 2\frac{39}{50}$$

### Approximate

Sometimes in shop work, particularly when working with inches, we need to know how many 16ths or 32nds of an inch we are working with. As we saw above, 2.78″ does not reduce to 16ths of an inch. But we can find out which 16th it is *closest* to $\left(2\dfrac{8}{16}, 2\dfrac{9}{16}, 2\dfrac{10}{16}, \text{etc.}\right)$.

Multiply the fractional part of the decimal number (0.78″ in the example above) by the desired fractional part (16 for 16ths, 32 for 32nds, etc.). Round the result to the nearest *whole* number. This is the top of your fraction.

***Example (1)*** Convert 2.78″ to the nearest 16th of an inch.

$$2.78'' = 2'' + 0.78''$$
$$\longrightarrow 0.78 \times 16 = 12.48$$
$$\text{or } 12 \text{ (rounded)}$$

Thus 2.78″ is approximately $2\dfrac{12}{16}$ or $2\dfrac{3}{4}$ (reduced).

---

**Note:** *2.78″ is not exactly $2\dfrac{12''}{16}$, but it is closer than $2\dfrac{11''}{16}$ or $2\dfrac{13''}{16}$.*

---

*Example (2)*   Convert 3.901″ to the nearest 32nd inch.

$$3.901″ = 3″ + 0.901″$$

$$\hookrightarrow 0.901 \times 32 = 28.832 \text{ or } 29$$

Thus 3.901 is approximately $3\dfrac{29″}{32}$ .

## Practice Exercise No. 16

Convert the following decimal fractions to the nearest 16th of an inch.

**(1)**   12.59″              **(2)**   4.79″              **(3)**   3.8125″

Convert the following decimal fractions to the nearest 64th of an inch.

**(4)**   7.9851″              **(5)**   12.51397″              **(6)**   2.7591″

---

Refer to *Calculator Hint* C.2 (Appendix C).

---

## 2.9   PRACTICAL APPLICATIONS EXERCISE NO. 3

**(1)**   A metal worker is asked to produce a bar 0.610″ in diameter. In the metal store, the following materials are available.

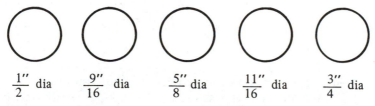

$\dfrac{1″}{2}$ dia        $\dfrac{9″}{16}$ dia        $\dfrac{5″}{8}$ dia        $\dfrac{11″}{16}$ dia        $\dfrac{3″}{4}$ dia

Which bar will be chosen if the least amount of material is to be wasted when machining diameter *D*? (See Fig. 2–9.) That is, which is closest to 0.610″ and at the same time larger than 0.610″? How far will the cutting tool have to be wound (or fed) to produce the required size? That is, what is the distance *X*? (*D* is the diameter of the bar chosen; see Fig. 2–9.)

**\*(2)**   A clerk receives two invoices for material. One firm sends two items having masses of 1.34 kg and 1.78 kg. The other firm sends three items having masses of $1\dfrac{1}{2}$ kg, $1\dfrac{5}{8}$ kg, and $1\dfrac{1}{5}$ kg. Calculate the total mass of the material received (give answer in decimal form).

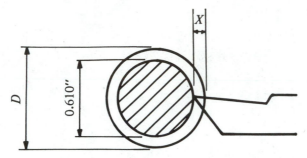

**Fig. 2–9**

(3) Work out the following problems, giving your answer in decimal form.

(a) $\dfrac{1}{2} + 0.382 + 1\dfrac{5}{8} + 0.633$

(b) $1.337 + 1\dfrac{3}{8} - 0.661$

(4) A circular plug fits into a flat plate as shown in Fig. 2–10. Calculate distances *A*, *B*, and *C* (answer in decimal form).

(5) Figure 2–11 shows a finished part as sketched by a fitter.

(a) What is the total length *L*? (Give your answer as a fraction of an inch.)

(b) If, when cutting the length, $\dfrac{3''}{64}$ extra must be allowed for machining, what is the length when first cut (i.e., what is $L + \dfrac{3''}{64}$)?

(6) You wish to cut a groove 0.649″ wide, but you haven't got a cutter this width and must choose one from Fig. 2–12. Which one should you choose?

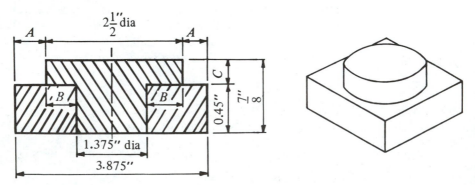

**Fig. 2–10**

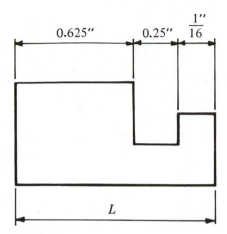

**Fig. 2–11**

(7)  You will have to make 2 cuts. When you have selected the cutter, you must take one cut and then move a distance $X$ to make a second cut. Calculate $X$, as shown in Fig. 2–13.

(8)  A person works $47\frac{1}{2}$ hours per week at $13.70 per hour. What is the gross weekly wage of this person?

*(9)  There are 17 rods being made, each 4.93 m long. If the rods are placed end to end, what is their total length?

*(10)  Using this table,

$$1" = 25.4 \text{ mm}$$
$$1 \text{ pt} = 0.568 \text{ liters } (\ell)$$
$$1 \text{ lb} = 0.454 \text{ kg}$$

convert the following to metric units.

(a)  14.5"      (b)  2'3"      (c)  2.5 pts      (d)  $3\frac{1}{2}$ gal

(e)  7.85 lb    (f)  8 oz      (g)  0.75"        (h)  3 yd

**Fig. 2–12**

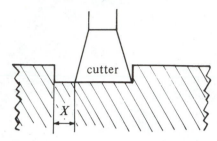

**Fig. 2–13**

**\*(11)** Given that 1 cm = 0.394″, convert the following to inches.

| | | | | |
|---|---|---|---|---|
| **(a)** 34 cm | **(b)** 59 cm | **(c)** 1 m | **(d)** 25.4 cm |
| **(e)** 800 cm | **(f)** 1.3 cm | **(g)** 0.75 cm | **(h)** 8 mm |

**(12)** If the average time allowed for a plumber and apprentice to install a complete wash basin is 2.85 hours, what would be the estimated labor cost of fitting 9 basins? These are the rates:
Plumber—$25 per hour
Apprentice—$15 per hour

**(13)** Another plumbing firm has these rates:
Plumber—$35.40 per hour
Apprentice—$18.00 per hour
Calculate this firm's estimated labor cost for the job described in Exercise 12.

## 2.10 PRACTICAL APPLICATIONS
### EXERCISE NO. 4

**\*(1)** A container holds components. The total weight of the container and components is 212 kg and the container weighs 59 kg. If the container holds 15 components, find the weight of each one. Each component weighs the same.

**(2)** A rod consists of 13 equal sections joined together. If the total length of rod is 12.96 yd, calculate the length of each section. (Answer to 3 decimal places.)

**(3)** A bill arrives for a quantity of tiles. The total charge for the tiles, priced at 18¢ each, is $47.52. Unfortunately, the quantity written on the bill cannot be read. Calculate this quantity.

**(4)** The sum $333.50 has to be evenly divided between 24 people. Calculate how much (to the nearest cent) each person receives.

**(5)** There are 15 small holes equally spaced around a large hole in a plate. (See Fig. 2–14.) Calculate the angle, θ, between two adjacent holes.

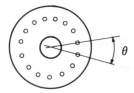

**Fig. 2–14**

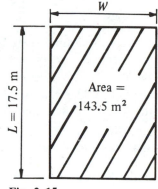

**Fig. 2–15**

*(6) A rectangular plate, shown in Fig. 2–15, has an area of 143.5 m². If area = length × width and the length is 17.5 m, calculate the width, W.

(7) A person has $20 to spend on items that are 68¢ each. How many can be bought?

(8) James Hunt covers 325 miles in 3.12 hours in the British Grand Prix. Calculate his average speed (correct to 2 decimal places).

## 2.11   MISCELLANEOUS EXERCISES

*(1) Two steel plates 10.35 mm and 18.75 mm thick are welded together. What is the total thickness of the plates after welding?

*(2) The overall thickness of a hardback textbook is 25 mm. If each of the two hardback covers is 1.4 mm thick, what is the total thickness of the pages between?

(3) The length of 115.2″ on a scale is divided into 12 equal parts. What length does *one* of the divisions represent?

*(4) A dimension of 2.5″ on a drawing has to be converted to millimeters. What will be the dimension in millimeters? (1″ = 25.4 mm.)

(5) Find the value of $\dfrac{72.6 \times 12.1}{24.2}$.

(6) Five pieces of plate each 5.6″ thick are placed one on top of the other. What is the total thickness of the pile?

(7) A drill 6″ long breaks 5.125″ from one end. What is the length of the remaining part of the drill?

(8) The length of a gauge is 1.2″. Express the length of the gauge in millimeters. (1″ = 25.4 mm.)

**(9)** Find the value of $\dfrac{20.25 \times 4.8}{40.5}$.

**\*(10)** Three pieces of chrome steel 14.3 mm, 15.9 mm, and 12.07 mm in length, respectively, are placed end to end in a straight line. Determine the total length.

**\*(11)** A center punch 72.5 mm long breaks 29.7 mm from one end. Determine the length of the remaining part of the center punch.

**(12)** A pair of dividers is set at 1.875″. What is the maximum number of times that this length can be used on a straight line 15″ long?

**(13)** Determine the value of $4\dfrac{1}{2} \times 25.4$.

**(14)** Determine the value of $\dfrac{51.3 \times 5.7}{17.1}$.

**(15)** Evaluate the following problems, without the use of mathematical tables, giving the answers correct to two places of decimals.
(a) $250.6 \times 0.35$
(b) $348.8 \div 4.2$

**(16)** A purchase of seven quart cans, six gallon cans, and eleven pint cans of paint has been made. Calculate, in quarts, the total amount of paint purchased.

**(17)** Calculate the total weight, in pounds, of the following 30 cans of paint.

| Size of Can | Quantity of Cans | Weight of Each Can |
|---|---|---|
| (a)   5 qt | 6 | 6.400 lb |
| (b)  0.5 qt | 12 | 0.590 lb |
| (c)  2.5 qt | 5 | 3.020 lb |
| (d)   1 qt | 7 | 1.060 lb |

**(18)** Calculate the cost of 16 cubic yards of ready mixed concrete at $450.25 per cubic yard.

**\*(19)** Calculate the cost of steel per kilogram if one metric ton costs $720.28. (1 metric ton = 1000 kilograms.)

**(20)** Change $\dfrac{5}{9}$ to a decimal fraction correct to three decimal places.

**(21)** Evaluate the following problem, without the use of mathematical tables, giving the answer correct to two decimal places.

$$\frac{3.85 \times 6.4}{2.9}$$

(22)  The points of a spark plug are to be set to a gap of 0.032″. Two feeler gauges are used to measure the setting. One gauge is 0.012″ thick. What is the thickness of the other gauge?

(23)  In Fig. 2–16, 0.250″ is to be removed from the workpiece by a milling cutter. The work table is raised by a calibrated wheel. Each revolution of the wheel raises the table 0.080″. How many revolutions of the wheel are required to raise the work sufficient to leave 0.010″ for a finishing cut?

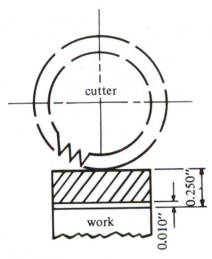

cutter

work

0.250″

0.010″

**Fig. 2–16**

# 3

# Approximation and Estimation

**Self-Assessment Test**

Write down the following numbers correct to the number of significant digits (abbreviated s.d.) stated.

(1)  16.751 to 3 s.d.              (2)  191.34 to 4 s.d.

(3)  6.751 to 3 s.d.               (4)  0.003 135 to 2 s.d.

(5)  171 000 to 1 s.d.

Obtain an approximate answer to the following calculations by correcting all the numbers to 1 s.d.

(6)  35.1 × 141                    (7)  1610 × 0.013

(8)  110 × 16                      (9)  1595 ÷ 41

(10)  9.65 ÷ 99.5

A craftsman asked to produce a component 1.312 456 789″ long would find it impossible since the degree of accuracy of this number is so high. The number must be rounded off to a figure that is possible to use in production work, that is, 1.3125″.

## 3.1  DECIMAL PLACES

One way to obtain approximate values is to state the number to so many **decimal places**. This has already been covered in Chapter 2, Section 2.7.

### Practice Exercise No. 1

Write down the following numbers correct to the number of decimal places stated.

| | | | | |
|---|---|---|---|---|
| **(1)** 5.239 771 | **(a)** to 5 | **(b)** to 3 | **(c)** to 2 | **(d)** to 1 |
| **(2)** 35.276 79 | **(a)** to 1 | **(b)** to 3 | **(c)** to 2 | **(d)** to 4 |
| **(3)** 0.004 997 | **(a)** to 5 | **(b)** to 4 | **(c)** to 3 | **(d)** to 1 |
| **(4)** 0.852 99 | **(a)** to 1 | **(b)** to 4 | **(c)** to 2 | **(d)** to 3 |
| **(5)** 110.973 251 | **(a)** to 4 | **(b)** to 3 | **(c)** to 1 | **(d)** to 2 |

## 3.2   SIGNIFICANT DIGITS

If a number is rounded off to "so many significant digits," the significant digits are determined irrespective of the position of the decimal point. We "count" significant digits by starting with the first nonzero digit (looking left to right). We then count that digit and all digits to the right of it, whether they are nonzero or zero. When we have counted the desired number of significant digits, we stop and round off the number at that position as we did in Section 2.7.

***Example (1)***   Correct 181.976 to 2 significant digits.

***Answer***   181.976 = 180 (to 2 s.d.)

With significant digits, zeros may have to be inserted to show the position of the decimal point.

***Example (2)***   Correct 161 350.7 to 2 significant digits.

***Answer***   161 350.7 = 160 000 (to 2 s.d.)

It must also be remembered that the first digit, *other than zero*, of a number is the first significant digit.

***Example (3)***   Correct the number 0.003 074 to 3 significant digits. 3 is the first significant digit, but the zero which *follows* the 3 is a significant digit also.

***Answer***   0.003 074 = 0.003 07 (to 3 s.d.)

***Example (4)***   Correct 0.003 102 5 to 3 significant digits.

***Answer***   0.003 102 5 = 0.003 10 (to 3 s.d.) Note that it is necessary in this example to retain the final 0.

***Example (5)***   Correct 100.371 to 4 significant digits.

***Answer***   100.371 = 100.4 (to 4 s.d.)

### Practice Exercise No. 2

Write down the following numbers correct to the number of significant digits stated.

| | | | | |
|---|---|---|---|---|
| **(1)** 371.8159 | **(a)** to 6 | **(b)** to 4 | **(c)** to 2 | **(d)** to 1 |
| **(2)** 0.018 625 | **(a)** to 4 | **(b)** to 3 | **(c)** to 2 | **(d)** to 1 |

(3) 44.713 65       **(a)** to 5     **(b)** to 6     **(c)** to 3     **(d)** to 2

(4) 16.981 271      **(a)** to 7     **(b)** to 2     **(c)** to 4     **(d)** to 1

(5) 0.006 173 91     **(a)** to 1     **(b)** to 3     **(c)** to 4     **(d)** to 5

## 3.3 ESTIMATION

It is very useful to know how to estimate answers to calculations. Even when using calculators, small errors occurring when inputting information may result in large errors in the final answer. It is therefore important to know how to quickly estimate the magnitude of an answer. This may be done by rounding off all the numbers involved to 1 significant digit.

***Example (1)*** Estimate the value of $20.9 \times 337$.

***Answer*** Rounding off all numbers to 1 s.d., estimated value $= 20 \times 300 = 6000$.

***Example (2)*** Estimate the value of $\dfrac{480 \times 639 \times 405}{7797}$.

***Answer*** Rounding off all numbers to 1 s.d.,

$$\text{estimated value} = \frac{500 \times \overset{30}{\cancel{600}} \times \overset{1}{\cancel{400}}}{\underset{\underset{1}{2}}{\cancel{8000}}} = 15\ 000.$$

Notice that we "canceled" by working with only one number from the top at a time.

$$\frac{500 \times 600 \times 400}{8000} \quad \text{became} \quad \frac{500 \times 600 \times 4}{80}$$

by dividing 400 and 8000 by 100. We then obtained

$$\frac{500 \times 60 \times 4}{8}$$

by dividing 600 and 80 by 10. Both of these reductions "looked" like we were crossing out zeros.

$$\frac{6\cancel{0}\cancel{0}}{8\cancel{0}}$$

which seems like a nice shortcut when dividing by 10, 100, etc. Then

$$\frac{500 \times 60 \times 4}{8} \quad \text{became} \quad \frac{500 \times 60 \times \overset{1}{\cancel{4}}}{\underset{2}{\cancel{8}}}$$

which became

$$\frac{500 \times \overset{30}{\cancel{60}} \times 1}{\underset{1}{\cancel{2}}} = \frac{500 \times 30 \times 1}{1}$$
$$= 15\ 000$$

**Example (3)**   Based on the example above, which of the following calculated results is most likely to be the answer to $\dfrac{480 \times 639 \times 405}{7797}$ ?

**(a)**   15 932.28
**(b)**   25 178.31
**(c)**   1595.2313

**Answer**   **(a)**; the estimate of 15 000 is closest to that number. Notice that the estimation can be quite different from the true answer (15 931.973). To truly compare them, both must be rounded to 1 significant digit.

## Practice Exercise No. 3

Obtain an approximate answer to the following calculations by correcting all the numbers to 1 significant digit (give answers to 1 s.d.).

> **Note:**   *Do not compute the exact answer!*

**(1)**   $10.1 \times 8.9$

**(2)**   $35.1 \times 810$

**(3)**   $77.9 \times 16.5$

**(4)**   $151 \times 0.937$

**(5)**   $29.6 \times 34.1$

**(6)**   $87.7 \times 9.2$

**(7)**   $0.003 \times 227.9$

**(8)**   $0.0884 \times 110$

**(9)**   $\dfrac{110.5}{0.75}$

**(10)**   $\dfrac{0.005\ 62}{130}$

**(11)**   $\dfrac{14.3 \times 15.9}{215}$

**(12)**   $\dfrac{97 \times 0.0125}{600}$

**(13)**   $\dfrac{84 \times 97 \times 15}{0.012}$

**(14)**   $\dfrac{61.5 \times 0.13}{79.6}$

**(15)**   $\dfrac{41.7 \times 88}{97}$

**(16)**   $\dfrac{1375 \times 97}{41.9}$

**(17)**   $\dfrac{1321 \times 14 \times 15.1}{77.9 \times 33}$

**(18)**   $\dfrac{997 \times 99 \times 101}{103 \times 12}$

**(19)**   $\dfrac{1\ 061\ 000 \times 15.5}{1065 \times 19.3}$

**(20)**   $\dfrac{16.23 \times 75.7 \times 16.9}{12.3 \times 98.4 \times 47.1}$

## 3.4   MISCELLANEOUS EXERCISES

**(1)**   Write the number 21.786
  **(a)**   Correct to 2 decimal places
  **(b)**   To the nearest whole number

**(2)**   Express 0.0488
  **(a)**   Correct to 2 decimal places
  **(b)**   Correct to 2 significant digits

**(3)**   Express 1.0554 mm correct
  **(a)**   To 2 decimal places
  **(b)**   To 3 significant digits

**(4)**   Express 2.758 13″ correct
  **(a)**   To the nearest whole number
  **(b)**   To 2 significant digits

**(5)**   Express 85.418 miles correct
  **(a)**   To the nearest mile
  **(b)**   To the nearest tenth of a mile

# 4

## Percents

---

*Self-Assessment Test No. 1*

Convert the following fractions to percents.

**(1)** $\frac{1}{4}$      **(2)** $\frac{1}{5}$      **(3)** $\frac{1}{8}$      **(4)** $\frac{5}{20}$      **(5)** $\frac{3}{5}$

Convert the following decimals to percents.

**(6)** 0.7      **(7)** 0.53      **(8)** 0.667      **(9)** 0.12      **(10)** 0.34

Convert the following percents to fractions.

**(11)** 45%      **(12)** 77%      **(13)** 95%      **(14)** 28%      **(15)** $8\frac{1}{2}\%$

Convert the following percents to decimals.

**(16)** 39%      **(17)** 64%      **(18)** 95%      **(19)** 75%      **(20)** 20%

A **percent** is a fraction with the denominator 100; for example,

$$20\% = \frac{20}{100}$$

which is equivalent to $\frac{2}{10}$ or $\frac{1}{5}$.

## 4.1 CONVERSIONS TO PERCENTS

### Conversion of Fractions to Percents

To convert a fraction to a percent, simply multiply by 100%.

*Example (1)*   Convert $\frac{4}{5}$ to a percent.

*Answer* $\dfrac{4}{5} = \dfrac{4}{\cancel{5}} \times \cancel{100}^{20}\% = 80\%$

$\phantom{Answer \dfrac{4}{5} = \dfrac{4}{5} \times 100}1$

*Example (2)* Convert $\dfrac{3}{8}$ to a percent.

*Answer* $\dfrac{3}{8} = \dfrac{3}{\cancel{8}} \times \cancel{100}^{25}\% = \dfrac{75}{2} = 37\dfrac{1}{2}\%$

$\phantom{Answer \dfrac{3}{8} = \dfrac{3}{8} \times}2$

---

**Note:** *The value of the fraction is not altered by multiplying by 100% since this is equivalent to multiplying by 1.*

---

## Practice Exercise No. 1

Convert the following fractions to percents.

| | | | | |
|---|---|---|---|---|
| **(1)** $\dfrac{1}{2}$ | **(2)** $\dfrac{1}{4}$ | **(3)** $\dfrac{1}{5}$ | **(4)** $\dfrac{9}{10}$ | **(5)** $\dfrac{5}{8}$ |
| **(6)** $\dfrac{7}{8}$ | **(7)** $\dfrac{1}{3}$ | **(8)** $\dfrac{2}{3}$ | **(9)** $\dfrac{1}{10}$ | **(10)** $\dfrac{3}{4}$ |
| **(11)** $\dfrac{5}{9}$ | **(12)** $\dfrac{11}{12}$ | **(13)** $\dfrac{4}{25}$ | **(14)** $\dfrac{9}{20}$ | **(15)** $\dfrac{1}{200}$ |
| **(16)** $\dfrac{7}{150}$ | **(17)** $\dfrac{3}{7}$ | **(18)** $\dfrac{1}{12}$ | | |

## Conversion of Decimals to Percents

To convert a decimal to a percent, simply multiply by 100%.

*Example (1)* $0.5 = 0.5 \times 100\% = 50\%$

*Example (2)* $0.21 = 0.21 \times 100\% = 21\%$

## Practice Exercise No. 2

Convert the following decimals to percents.

| | | | | |
|---|---|---|---|---|
| **(1)** 0.60 | **(2)** 0.87 | **(3)** 0.875 | **(4)** 0.143 | **(5)** 0.075 |
| **(6)** 0.003 | **(7)** 0.955 | **(8)** 0.53 | **(9)** 0.625 | **(10)** 0.135 |
| **(11)** 0.002 | **(12)** 0.413 | **(13)** 0.0621 | **(14)** 0.81 | **(15)** 0.719 |
| **(16)** 0.388 | **(17)** 0.157 | **(18)** 0.897 | | |

## 4.2 CONVERSIONS FROM PERCENTS

### Conversion of Percents to Fractions

To convert a percent to a fraction, simply divide the percent by 100% and cancel to the lowest terms.

*Example (1)*   $80\% = \dfrac{\overset{4}{\cancel{80}}}{\underset{5}{\cancel{100}}} = \dfrac{4}{5}$

*Example (2)*   $35\% = \dfrac{\overset{7}{\cancel{35}}}{\underset{20}{\cancel{100}}} = \dfrac{7}{20}$

*Example (3)*   $2\frac{1}{2}\% = \dfrac{2\frac{1}{2}}{100} = \dfrac{\left(\frac{5}{2}\right)}{\left(\frac{100}{1}\right)} = \dfrac{\overset{1}{\cancel{5}}}{2} \times \dfrac{1}{\underset{20}{\cancel{100}}} = \dfrac{1}{40}$

### Practice Exercise No. 3

Convert the following percents to fractions.

(1)  30%    (2)  18%    (3)  88%    (4)  12%    (5)  $12\frac{1}{2}\%$

(6)  14%    (7)  39%    (8)  66%    (9)  48%    (10)  $33\frac{1}{3}\%$

(11)  $66\frac{2}{3}\%$    (12)  $7\frac{1}{2}\%$    (13)  13%    (14)  $37\frac{1}{2}\%$    (15)  95%

(16)  54%    (17)  $87\frac{1}{2}\%$    (18)  $8\frac{1}{2}\%$

### Conversion of Percents to Decimals

To convert a percent to a decimal, simply divide the percent by 100%.

*Example (1)*   $80\% = \dfrac{80}{100} = 0.8$

*Example (2)*   $13.5\% = \dfrac{13.5}{100} = 0.135$

*Example (3)*   $13\frac{1}{4}\% = 13.25\% = \dfrac{13.25}{100} = 0.1325$

## Practice Exercise No. 4

Convert the following percents to decimals.

(1)  37%     (2)  49%     (3)  18.5%     (4)  16.8%     (5)  $5\frac{1}{4}\%$

(6)  $9\frac{1}{2}\%$     (7)  $13\frac{1}{8}\%$     (8)  $43\frac{1}{4}\%$     (9)  $12\frac{1}{2}\%$     (10)  98%

(11)  $44\frac{1}{2}\%$     (12)  $2\frac{1}{2}\%$     (13)  $3\frac{3}{8}\%$     (14)  $4\frac{1}{4}\%$     (15)  $55\frac{1}{2}\%$

---

*Self-Assessment Test No. 2*

Find the value of:

(1)  5% of $600        (2)  16% of $800        (3)  4% of $99

*(4)  $5\frac{1}{2}\%$ of 1600 mm       (5)  22% of 160 lb

Express the first of each of the following pairs of numbers as a percent of the second.

(6)  45, 90            (7)  16, 64            *(8)  250 g, 2000 g

(9)  5¢, $2            (10)  25 lb, 2 tons

---

## 4.3  SOLVING PROBLEMS WITH PERCENTS

There are three common types of questions encountered when using percents.

### Type A

*Example (1)*  Find 5% of $300.

*Method*  Convert 5% to a fraction; that is,

$$5\% = \frac{5}{100} = \frac{1}{20}$$

Remember that "of" usually means *times* ($\times$).

$$\therefore 5\% \text{ of } \$300 = \frac{5}{\cancel{100}} \times \$3\cancel{00} = \$15$$

*Example (2)*  Find 7% of $420.

$$7\% \text{ of } \$420 = \frac{7}{\cancel{100}} \times \$42\cancel{0} = \frac{\$294}{10} = \$29.40$$

*Example (3)*  Find $9\frac{1}{2}\%$ of $750.

$$9\frac{1}{2}\% \text{ of } \$750 = \frac{9\frac{1}{2}}{100} \times \$750 = \frac{\$\frac{19}{2}}{\frac{10}{1}} \times 75$$

$$= \frac{\$19}{2} \times \frac{1}{\underset{2}{10}} \times \overset{15}{75} = \frac{\$285}{4} = \$71.25$$

## Type B

*Example (1)*  Express 20 as a percent of 80.

*Method*  Express 20 as a fraction of 80; that is,

$$\frac{20}{80}$$

Convert to a percent by multiplying by 100%.

$$\therefore 20 \text{ as a percent of } 80 = \frac{\overset{1}{20}}{\underset{4}{80}} \times 100\%$$

$$= \frac{100\%}{4} = 25\%$$

*Example (2)*  What percent of $4.50 is 15¢?

*Answer*  $\dfrac{\overset{1}{15}}{\underset{3}{450}} \times 100\% = \dfrac{10\%}{3} = 3\frac{1}{3}\%$

Notice the number following the word "of" goes on the bottom.

## Type C

*Example (1)*  5% of what number is $15?

*Method*  Express as a fraction.

$$\frac{\$15}{5\%}$$

Notice the "%" goes on the bottom. (This is similar to Type B, where you find the "%" number using the "of" number. In Type C we find the "of" number using the "%" number.) Now express 5% as a decimal and divide.

*Answer*  $\dfrac{\$15}{5\%} = \dfrac{\$15}{0.05} = \$300$

*Example (2)*  30% of Ben's salary goes to taxes. The taxes are \$127.50. What is Ben's salary (before taxes)? Rewrite as: 30% of the salary is \$127.50

*Answer*  Salary $= \dfrac{\$127.50}{30\%} = \dfrac{\$127.50}{0.30} = \$425.00$

## Practice Exercise No. 5

Find the value of:

(1)  3% of \$420

(2)  8% of \$900

(3)  12% of \$450

(4)  4% of 220 miles

(5)  30% of 65 pints

(6)  125% of 600 lb

*(7)  38% of 1 km

*(8)  40% of 9.3 kg

*(9)  16% of 2.12 m

(10)  63% of \$4500

(11)  $7\frac{1}{2}\%$ of \$8000

(12)  19% of 25.5″

## Practice Exercise No. 6

Express the first of each of the following pairs of quantities as a percent of the second.

(1)  3, 8    (2)  \$15, \$60    *(3)  18 m, 36 m    (4)  5¢, \$1    (5)  200 oz, 1 lb

(6)  3″, 3 ft    (7)  13, 32    (8)  3, 24    (9)  16, 160    (10)  $123\frac{1}{2}$ ft, 50 ft

## Practice Exercise No. 7

Find the missing numbers.

(1)  15% of ____ is 48.

(2)  12% of ____ is 5400.

(3)  2.3% of ____ is 1.035.

(4)  $\frac{1}{2}\%$ of ____ is 25 mg.

## 4.4  PRACTICAL APPLICATIONS
### EXERCISE NO. 1

(1)  It takes 40 minutes to machine a certain component using a high-speed steel cutter. It is claimed that, using a tungsten carbide cutter, the time for the job could be reduced by 35%. What would be the new time for the job?

(2)  A machinist produces a batch of components. Out of 4800 produced, 96 were scrapped. Find the percent scrapped.

**\*(3)** When casting a component, allowance has to be made for shrinkage of metal. If the final dimension on a cast part is to be 3.75 m (when cool), and 2% is allowed extra on the pattern to cover shrinkage, what will this dimension be when the metal is hot?

**(4)** Solder used by plumbers contains 30% tin, 69% lead, and 1% antimony. How much of each is contained in 400 ounces of solder?

**(5)** A woman earning $28 000 per year is given a 5% increase. Calculate her new wage.

**(6)** Owing to a shortage of orders, a firm must reduce its staff by 15%. If it presently employs 1750 people
**(a)** How many people will be laid-off?
**(b)** How many people will still be employed?

**(7)** A car salesman receives a basic wage of $450 per week plus 1% commission on all sales.
**(a)** Calculate his commission.
**(b)** Calculate his total wage for the following weeks.

| Week | Total Sales |
|------|-------------|
| 1 | $37 500 |
| 2 | $23 500 |
| 3 | $59 700 |
| 4 | $140 000 |
| 5 | $5750 |

**(8)** A person pays 11% interest on a loan of $550. How much interest is paid?

**(9)** A shopkeeper buys goods at 55¢ each and wants to make a profit of 20% on each item sold. Calculate the selling price.

**(10)** A shopkeeper buys goods at $1.28 each and wants to make a profit of $12\frac{1}{2}\%$ on each item sold. Calculate the selling price.

**(11)** A machine shop job requires 35 hours on the lathe, 7 hours on the milling machine, and 8 hours on the drill press. What percent of the total time should be charged to each machine?

**\*(12)** A component is to be produced as shown in Fig. 4–1. Instructions indicate that all dimensions should have a tolerance of ±1%. Work out the actual limits of each dimension.

**(13)** A buying department purchases goods from a company. If the department buys a lot of goods, it gets the following discount.

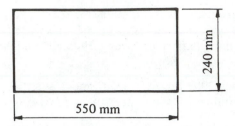

**Fig. 4–1**

| Value of Goods Bought | Discount (%) |
|---|---|
| $0    –$999 | 0 |
| $1000  –$4999 | 3 |
| $5000  –$9999 | 5 |
| $10 000 and over | 10 |

    **(a)** Calculate the discount.

    **(b)** Calculate the actual cost of goods valued at $333, $3950, $12 500, and $7750.

**(14)** A certain building job is estimated to need a total of 8500 bricks plus an allowance of 8% for cutting and waste. How many bricks should be ordered?

---

*Self-Assessment Test No. 3*

  **(1)** The price of an article is increased from $1 to $1.50. Calculate this as a percent increase.

  **(2)** The price of an article is reduced from $2 to $1.50. Calculate this as a percent decrease.

  **(3)** An article is bought for 8¢ and sold for 10¢. What is the percent profit?

  **(4)** A metal rod expands when heated from 4″ to 4.01″. What is the percent increase in length?

 **\*(5)** A container holds 2000 cm$^3$ of liquid. It leaks and the level falls to 1700 cm$^3$. Calculate the percent loss of fluid.

---

## 4.5  PERCENT GAIN AND LOSS

When dealing with problems involving percent gain or loss, it is important to express the gain or loss as a fraction of the **original value** before converting to a percent.

*Example (1)*  A bar 350 mm long is heated and extends to 355 mm long. Find the percent increase in length.

*Answer*   % increase $= \dfrac{\text{increase}}{\text{original length}} \times 100\%$

$$= \dfrac{\overset{1}{\cancel{5}}}{\underset{7}{\cancel{350}}} \times \cancel{100}\% = 1\tfrac{3}{7}\%$$

*Example (2)*   A person buys a car for $4500 and sells it for $3000. Calculate the percent loss made.

*Answer*   % loss $= \dfrac{\text{loss}}{\text{original price}} \times 100\%$

$$= \dfrac{\overset{1}{\cancel{1500}}}{\underset{3}{\cancel{4500}}} \times 100\% = 33\tfrac{1}{3}\%$$

## 4.6   PRACTICAL APPLICATIONS
## EXERCISE NO. 2

(1)   A worker's wage is increased from $150 per week to $180 per week. Calculate this as a percent increase.

*(2)   The correct answer to a problem is 120 mm; Roger arrives at the value 132 mm. Calculate his percent error.

(3)   A machine pulley running free turns at 450 rev/min. When the machine is loaded, the speed drops to 430 rev/min. Find the percent drop in speed.

(4)   The bore of a water pipe is 20″ in diameter. The pipe becomes coated to a depth of 1″ with scale. Calculate the percent reduction in its diameter.

(5)   A garment is purchased for $7.50 and is marked up for sale at $9.00. Calculate the percent profit.

(6)   The estimated cost for a building was $600 000. The following year this cost had risen to $750 000. Calculate the percent rise in cost.

*(7)   A mass is attached to a spring 300 mm long. The spring then extends to 350 mm. Calculate the percent increase in length.

(8)   A craftsman machines a component on a lathe in 25 minutes. Using a new type of cutting tool, this time is reduced to 22½ minutes. Calculate the percent reduction in time.

## 4.7 MISCELLANEOUS EXERCISES

(1) A metal is composed of 70% copper and 30% zinc. What mass of copper is contained in a block of metal that contains 60 lb of zinc?

*(2) When 16% full, a tank contains 5.7 liters. What will it contain when 35% full? Give the answer correct to two places of decimals.

(3) A filter has a normal selling price of $2. What would trade customers pay for it if their trade discount of 25% is deducted from this price and a tax of 8% of the trade price is then added?

(4) A weaver receives $15.98 per hour, and a doffer receives $5.95 per hour. Express the wage of the doffer as a percentage of that of the weaver. Answer correct to two decimal places.

(5) A price reduction of 10% is offered for cash on a sewing machine marked at $110. What is the cash price of the sewing machine?

(6) How much is 8% of $625?

(7) A new car costs $8000. In its first year it loses 25% of its value at the beginning of the year. Every year after the first it loses another 15% of its value at the start of the year. Calculate its value after **(a)** 1 year, **(b)** 2 years, and **(c)** 3 years, to the nearest dollar.

(8) A planing machine working at the rate of 40 strokes per minute increases its speed by 15%. Find the increase in the number of strokes per minute.

(9) Two castings are found to be defective in a total quantity of 50 manufactured. Express the number of defective castings as a percent of the total quantity manufactured.

*(10) A casting of mass 125 kg has 2% of its mass removed by machining. Determine the mass of metal removed.

(11) In a factory three machines are used to produce a component. If each machine produces 6000 components, of which 7%, 3%, and 2% are faulty, respectively, calculate the percent of faulty items in the total output from the factory.

*(12) A casting has a mass of 10 kg that is reduced by TWO successive machining operations. The first operation removes 10% of the original mass and the second removes 7% of the remaining mass. Calculate the finished mass of the casting.

# PART II

## Applied Algebra

# 5

# Fundamentals of Algebra

There are many types of practical problems that can best be solved by the use of formulas. Formulas are examples of **applied** algebra. To be proficient at using formulas, you need to know about equations and some of the fundamental skills of algebra.

This chapter covers some of those fundamental skills that are used in working with formulas and equations (these are covered in Chapter 6). Although there are only a few applications shown in this chapter, you will find many in Chapter 6.

---

**Self-Assessment Test No. 1**

Simplify the following.

(1) $a^3 \times a^2$

(2) $b^4 \times b^5 \times b^6$

(3) $a^2b^2 \times a^3b^3$

(4) $p^2q^2 \times p^4q^2r^3$

(5) $\dfrac{a^5}{a^2}$

(6) $\dfrac{a^2b^4}{b^5}$

(7) $\dfrac{x^2y^2z}{xyz}$

(8) $\dfrac{p^3q^3r^4}{q^2r^2}$

(9) $(x^3)^2$

(10) $(p^2q^2)^4$

(11) $\left(\dfrac{1}{3}x^2y\right)^4$

(12) $16^{1/2}$

(13) $8^{1/3}$

(14) $27^{1/3}$

(15) $20^0$

(16) $(ab)^0$

---

## 5.1 EXPONENTS

### What Are Exponents?

*Answer*   The quantity $3 \times 3 \times 3 \times 3 \times 3$ can be written $3^5$. This is called the fifth power of 3. The small figure 5 gives the number of 3's to be multiplied together and is called the **exponent**.

## The Laws of Exponents

### Multiplication

$$x^5 \cdot x^6 = (x \cdot x \cdot x \cdot x \cdot x) \cdot (x \cdot x \cdot x \cdot x \cdot x \cdot x) = x^{11}$$

That is,

$$x^5 \cdot x^6 = x^{5+6} = x^{11}$$

When two powers of the same letter or number are multiplied together, add the exponents.

**Example (1)**   $a^7 \times a = a^{7+1} = a^8$

**Example (2)**   $b^2 \times b^3 = b^{2+3} = b^5$

**Example (3)**   $c^3 \times c^2 \times c^4 = c^{3+2+4} = c^9$

**Example (4)**   $a^2 \times b^3 \times a^4 \times b^2 \times c = a^{2+4} \times b^{3+2} \times c^1$
$$= a^6 b^5 c$$

---

**Note:**   $c^1 = c.$

---

### Practice Exercise No. 1

Simplify the following.

(1)   $b^2 \times b^4$  

(2)   $b^3 \times b^5 \times b^9$  

(3)   $z^3 \times z^4 \times z^5$

(4)   $a^2 b^2 c^2 \times b^2 c^2 d^2$  

(5)   $a^4 b^2 c^3 \times a^2 bc$  

(6)   $p^2 r^2 t^2 \times p r^2 t$

(7)   $4^2 \times 4^2 \times 4^3$  

(8)   $9^2 \times 9 \times 9^5$  

(9)   $2^2 \times 2 \times 2^4$

(10)   $a^2 b^2 \times p^2 r^2 \times a^3 r^5$  

(11)   $10^2 \times 10^3 \times 10^4$  

(12)   $a^2 b^2 \times ab$

(13)   $M^2 N^2 O^2 \times MN^3$  

(14)   $A^3 B^3 \times B^4 C^2$  

(15)   $3A^2 B^3 \times 2AB^4$

(16)   $\dfrac{3}{4} A^4 B \times \dfrac{1}{4} A^2 B^2$

### Division

$$\frac{x^6}{x^4} = \frac{\overset{1}{\cancel{x}} \cdot \overset{1}{\cancel{x}} \cdot \overset{1}{\cancel{x}} \cdot \overset{1}{\cancel{x}} \cdot x \cdot x}{\underset{1}{\cancel{x}} \cdot \underset{1}{\cancel{x}} \cdot \underset{1}{\cancel{x}} \cdot \underset{1}{\cancel{x}}} = x^2$$

That is,

$$\frac{x^6}{x^4} = x^{6-4} = x^2$$

When dividing powers of the *same* letter or number, *subtract* the exponent of the denominator (bottom number) from the exponent of the numerator (top number).

**Example (1)**   $\dfrac{a^5}{a^2} = a^{5-2} = a^3$

**Example (2)**   $\dfrac{M^9}{M^4} = M^{9-4} = M^5$

**Example (3)**   $\dfrac{M^6N^3}{M^4N} = (M^{6-4})(N^{3-1}) = M^2N^2$

**Example (4)**   $\dfrac{P^4Q^3R^3}{PQ^2R^2} = (P^{4-1})(Q^{3-2})(R^{3-2}) = P^3QR$

## Practice Exercise No. 2

Simplify the following.

**(1)** $\dfrac{z^4}{z^3}$  **(2)** $\dfrac{a^8}{a^3}$  **(3)** $\dfrac{b^9}{b^5}$

**(4)** $\dfrac{c^{10}}{c^6}$  **(5)** $\dfrac{a^4b^5c^2}{a^2b}$  **(6)** $\dfrac{x^{14}y^{15}z}{x^2y^3}$

**(7)** $\dfrac{x^4y^5}{xy^2}$  **(8)** $\dfrac{4x^3y^4z^5}{2xyz^3}$  **(9)** $\dfrac{(3z^2y^3x^{14})(3zx^4y^9)}{9z^2x^5y^8}$

**(10)** $\dfrac{(13a^4b^5c^6)(2a^5b^3c)}{a^2b^2c^2}$

## Powers of a Power

$$(z^3)^5 = z^3 \times z^3 \times z^3 \times z^3 \times z^3 = z^{3+3+3+3+3} = z^{15}$$

That is,

$$(z^3)^5 = z^{3 \times 5} = z^{15}$$

When raising the power of a letter or number to a power, multiply the exponents together.

**Example (1)**   $(a^4)^3 = a^{4 \times 3} = a^{12}$

**Example (2)**   $(b^5)^6 = b^{5 \times 6} = b^{30}$

**Example (3)**   $(x^2y^2)^3 = (x^{2 \times 3})(y^{2 \times 3}) = x^6y^6$

**Example (4)**   $(a^4b^4c^2)^3 = (a^{4 \times 3})(b^{4 \times 3})(c^{2 \times 3}) = (a^{12})(b^{12})(c^6) = a^{12}b^{12}c^6$

**Example (5)**   $\left(\dfrac{2a^2b^2}{3xy}\right)^5 = \dfrac{(2^{1 \times 5})(a^{2 \times 5})(b^{2 \times 5})}{(3^{1 \times 5})(x^{1 \times 5})(y^{1 \times 5})} = \dfrac{2^5a^{10}b^{10}}{3^5x^5y^5}$

## Practice Exercise No. 3

Simplify the following.

(1)  $(x^2)^4$   (2)  $(x^9)^4$   (3)  $(z^4)^5$

(4)  $(x^3y^3z)^4$   (5)  $(x^9y^9z)^2$   (6)  $(ab^4c^3)^4$

(7)  $(x^3z^2)^{10}$   (8)  $\left(\dfrac{3M^4N^5}{2MN}\right)^3$   (9)  $\left(\dfrac{4a^4b^9}{3ab^5}\right)^2$

(10)  $\left(\dfrac{7a^4b^3c}{2a^3b^2}\right)^3$   (11)  $\left(\dfrac{x^4y^4}{a^2b^3}\right)^2$   (12)  $\left(\dfrac{2xy^4}{4x^3yz^2}\right)^3$

## Negative Exponents

$$\frac{M^5}{M^7} = \frac{\overset{1}{\cancel{M}} \times \overset{1}{\cancel{M}} \times \overset{1}{\cancel{M}} \times \overset{1}{\cancel{M}} \times \overset{1}{\cancel{M}}}{\underset{1}{\cancel{M}} \times \underset{1}{\cancel{M}} \times \underset{1}{\cancel{M}} \times \underset{1}{\cancel{M}} \times \underset{1}{\cancel{M}} \times M \times M} = \frac{1}{M^2}$$

But

$$\frac{M^5}{M^7} = M^{5-7} = M^{-2}$$

$$\therefore \frac{1}{M^2} = M^{-2}$$

A negative exponent represents a **reciprocal**.

*Example (1)*   $10^{-2} = \dfrac{1}{10^2} = \dfrac{1}{100}$

*Example (2)*   $a^{-3} = \dfrac{1}{a^3}$

*Example (3)*   $a^{-2}b^{-2} = \dfrac{1}{a^2b^2}$

*Example (4)*   $a^2b^{-3} = \dfrac{a^2}{b^3}$

*Example (5)*   $a^2b^{-4}c^{-3} = \dfrac{a^2}{b^4c^3}$

*Example (6)*   $\dfrac{1}{a^{-1}} = \dfrac{1}{\left(\dfrac{1}{a}\right)} = \dfrac{1}{1} \div \dfrac{1}{a} = \dfrac{1}{1} \times \dfrac{a}{1} = a$

## Practice Exercise No. 4

Rewrite the following giving answers without using negative exponents.

**(1)**  $z^{-4}$

**(2)**  $c^{-2}d^{-3}e^{-4}$

**(3)**  $a^2b^{-4}$

**(4)**  $x^{-3}y^{-4}z^5$

**(5)**  $x^{-3}y^4z^{-5}$

**(6)**  $a^4b^3c^{-3}$

**(7)**  $\dfrac{1}{b^{-4}}$

**(8)**  $\dfrac{a^2c^2}{b^{-2}}$

**(9)**  $\dfrac{3M^{-1}N^{-1}}{2p^{-3}q^{-4}}$

**(10)**  $\dfrac{6x^{-3}y^2}{2x^2y^{-3}}$

## Fractional Exponents

$a^{1/2}$ means the square root of $a$ (or $\sqrt{a}$)

$a^{1/3}$ means the cube root of $a$   (or $\sqrt[3]{a}$)

***Example (1)***   $4^{1/2} = 2$        Because $2 \times 2 = 4$

***Example (2)***   $64^{1/3} = 4$        Because $4 \times 4 \times 4 = 64$

***Example (3)***   $125^{1/3} = 5$        Because $5 \times 5 \times 5 = 125$

***Example (4)***   $6.25^{1/2} = 2.5$        Because $2.5 \times 2.5 = 6.25$

## Practice Exercise No. 5

Work out the following.

**(1)**  $49^{1/2}$

**(2)**  $81^{1/2}$

**(3)**  $27^{1/3}$

**(4)**  $64^{1/2}$

**(5)**  $1^{1/2}$

**(6)**  $1^{1/3}$

**(7)**  $144^{1/2}$

**(8)**  $100^{1/2}$

**(9)**  $1000^{1/3}$

**(10)**  $8^{1/3}$

**(11)**  $16^{1/2}$

**(12)**  $25^{1/2}$

## Zero Exponents

$$\frac{b^4}{b^4} = \frac{\overset{1}{\cancel{b}} \times \overset{1}{\cancel{b}} \times \overset{1}{\cancel{b}} \times \overset{1}{\cancel{b}}}{\underset{1}{\cancel{b}} \times \underset{1}{\cancel{b}} \times \underset{1}{\cancel{b}} \times \underset{1}{\cancel{b}}} = \frac{1}{1} = 1$$

But

$$\frac{b^4}{b^4} = b^{4-4} = b^0$$

$$\therefore b^0 = 1$$

Any number or letter raised to the power 0 is equal to 1.

***Example (1)***   $3^0 = 1$

***Example (2)*** $b^0 = 1$

***Example (3)*** $2^2 \times a^0 = 2^2 \times 1 = 4$

***Example (4)*** $3a^0b^2 = 3 \times 1 \times b^2 = 3b^2$

## Practice Exercise No. 6

Simplify the following.

**(1)** $(a^2b^3)(a^4b^4)(a^3b)$

**(2)** $\dfrac{(a^5bc^3)(a^4b^2c)}{ab^2c}$

**(3)** $(a^{-2}bc^2)(a^4b^3c^0)$

**(4)** $121^{1/2}$

**(5)** $a^0b^2c^0$

**(6)** $(16^{1/2})(8^{1/3})(4^{1/2})$

**(7)** $(a^4)^3 \times (a^5)^2 \times (a^4)^{1/2}$

**(8)** $\dfrac{(27^{1/3})(a^5)(b^5)}{4^{1/2}a^{-2}b^{-3}}$

**(9)** $\dfrac{(x^2y^4z^3)(x^4y^5z^6)}{x^3y^5z^{12}}$

**(10)** $\dfrac{9x^4y^4}{3x^{-3}y^{-2}}$

---

***Self-Assessment Test No. 2***

Work out the following.

**(1)** $(+2) + (+3)$  **(2)** $(+2) - (+3)$  **(3)** $(-2) + (+3)$  **(4)** $(-2) - (-5)$

**(5)** $(-1) - (+1)$  **(6)** $(+6) - (-2)$  **(7)** $(-5) + (+4)$  **(8)** $(-3) \times (-2)$

**(9)** $(-1) \times (+12)$ **(10)** $(-3) \div (+1)$ **(11)** $(+6) \div (+2)$ **(12)** $(-10) \div (-5)$

---

## 5.2 DIRECTED NUMBERS

Consider a thermometer. Temperatures may be recorded as $-3°C$ or $+5°C$. The $+$ and $-$ do not refer to addition or subtraction; $+5$ indicates a point $5°$ above the zero mark and $-3$ indicates a point $3°$ below it. Positive and negative numbers are used to distinguish between opposite qualities; for example, if the hours of the day are positive from noon, 5:00 P.M. would be written $+5$ whereas 9:00 A.M. would be written $-3$.

You must remember that if a number has no sign in front of it, it is assumed to be positive.

### Addition and Subtraction

When adding or subtracting directed numbers, a number line may be used.

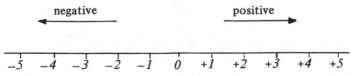

Movements to the right are positive and movements to the left are negative.

*Example (1)*   $(+3) - (+2) = (+3) - 2$

Start at 3 on the number line; $-2$ means move two places to the left.

*Answer*   $(+3) - (+2) = +1$

*Example (2)*   $(-3) + (+2) = (-3) + 2$

Start at $-3$ on the number line; $+2$ means move two places to the right.

*Answer*   $(-3) + (+2) = -1$

Two important facts to remember are:

**(1)**  When in an expression $+(-a)$ or $-(+a)$ appears, it may be written as $-a$, that is, if two signs appear *between* terms and they are different (one positive and one negative), then the result is a negative. For example,

$$3 - (+2) = 3 - 2 = 1$$
$$2 + (-3) = 2 - 3 = -1$$

**(2)**  When in an expression $+(+a)$ or $-(-a)$ appears, it may be written as $+a$, that is, if two signs appear *between* terms and they are the same (both positive or both negative), then the result is a positive. For example,

$$2 + (+3) = 2 + 3 \quad = 5$$
$$2 - (-3) = 2 + 3 \quad = 5$$
$$-2 - (-5) = -2 + 5 = 3$$
$$-2 + (+7) = -2 + 7 = 5$$

### Practice Exercise No. 7

Work out the following.

**(1)** $(+3) + (+2)$  **(2)** $(-3) - (+2)$  **(3)** $(+4) + (+3)$  **(4)** $(+2) - (+1)$

**(5)** $(-3) - (-2)$  **(6)** $(-5) + (+1)$  **(7)** $(+6) + (-1)$  **(8)** $(-5) - (-2)$

**(9)** $(+7) + (-1)$  **(10)** $(+3) - (+3)$  **(11)** $(+3) + (-2)$  **(12)** $(+13) - (-5)$

**(13)** $(-3) - (+1)$  **(14)** $(+6) + (+4)$  **(15)** $(-4) - (-2)$  **(16)** $(-5) + (-1)$

**(17)** $(+5) + (+3)$  **(18)** $(+17) - (-3)$ **(19)** $(+4) - (+2)$  **(20)** $(+1) + (-3)$

## Multiplication

When multiplying directed numbers, the numerical value of the product is found by multiplying in the ordinary way.

If both terms have like signs, that is, both $(+)$ or both $(-)$, then the product is positive. If the terms have unlike signs, that is, one $(+)$ and one $(-)$, then the product is negative. For example,

$$-2 \times -3 = +6 \qquad \text{signs the same} \therefore \text{product } (+)$$
$$-3 \times +1 = -3 \qquad \text{signs different} \therefore \text{product } (-)$$
$$+3 \times +4 = +12 \qquad \text{signs the same} \therefore \text{product } (+)$$
$$+2 \times -1 = -2 \qquad \text{signs different} \therefore \text{product } (-)$$

## Division

When dividing one number by another, the numerical value of the quotient is found by dividing in the ordinary way.

If both terms have like signs, that is, both $(+)$ or both $(-)$, then the quotient is positive. If both terms have unlike signs, that is, one $(+)$ and one $(-)$, then the quotient is negative. For example,

$$(+6) \div (+3) = \frac{+6}{+3} = +2 \qquad \text{signs the same} \therefore \text{quotient } (+)$$

$$(-3) \div (-1) = \frac{-3}{-1} = +3 \qquad \text{signs the same} \therefore \text{quotient } (+)$$

$$(-4) \div (+2) = \frac{-4}{+2} = -2 \qquad \text{signs different} \therefore \text{quotient } (-)$$

$$(+8) \div (-1) = \frac{+8}{-1} = -8 \qquad \text{signs different} \therefore \text{quotient } (-)$$

### Practice Exercise No. 8

Work out the following.

| | | |
|---|---|---|
| **(1)** $(+3) \times (+2)$ | **(2)** $(-3) \times (+1)$ | **(3)** $(+4) \times (-3)$ |
| **(4)** $(-3) \times (-2)$ | **(5)** $(-1) \times (+3)$ | **(6)** $(+10) \times (-5)$ |
| **(7)** $(+5) \times (+6)$ | **(8)** $(-4) \times (-5)$ | **(9)** $(-3) \times (-3)$ |
| **(10)** $(+20) \times (+3)$ | **(11)** $(-4) \times (-4)$ | **(12)** $(+3) \times (-15)$ |
| **(13)** $(-3) \div (+1)$ | **(14)** $(-2) \div (-2)$ | **(15)** $(+3) \div (+2)$ |
| **(16)** $(-14) \div (-2)$ | **(17)** $(+8) \div (-2)$ | **(18)** $(+8) \div (+4)$ |
| **(19)** $(+21) \div (-7)$ | **(20)** $(-24) \div (-8)$ | **(21)** $(+2) \div (+4)$ |
| **(22)** $(-4) \div (-1)$ | **(23)** $(-28) \div (+7)$ | **(24)** $(+50) \div (+10)$ |

---

**Self-Assessment Test No. 3**

Simplify the following by collecting like terms.

(1)   $3a + 2a + 5a$  (2)   $3x - 2y + 2x + 5y$

(3)   $z + 7z - 2q + z$  (4)   $p^2 + 2p - p + 4p^2$

(5)   $8a^2 - 2a + 14a - a^2$  (6)   $6ab + 7ba + 6b$

(7)   $8xy - 9y + 10yx$  (8)   $16x^3 - 14x^2 + 9x^3 - x^2$

---

## 5.3   COLLECTION OF LIKE TERMS

A **term** is a combination of a letter and a number, such as $6x$ or $3y^2$, and represents a multiplication of values. In these cases, $6x$ means "6 times the value of $x$," and $3y^2$ means "3 times $y$ squared."

 If you had 5 apples in a bag and then added 2 more apples, you would not say "I have 5 apples plus 2 apples"; you would say "I have 7 apples," that is, $5a + 2a = 7a$. You have collected together the like quantities.

***Example (1)***   Simplify $7x + 3y + 2x + 4y$.
Consider the terms in $x$,

$$7x + 2x = 9x$$

Consider the terms in $y$,

$$3y + 4y = 7y$$
$$\therefore 7x + 3y + 2x + 4y = 9x + 7y$$

***Example (2)***   Simplify $3x^2 + 2x - 2x^2 + x$.

---

**Note:**   *Terms in x and x² are not like terms; that is, terms are "like" only if their letter factors are identical, same letter and same exponent.*

---

Consider the terms in $x^2$,

$$3x^2 - 2x^2 = x^2$$

Consider the terms in $x$,

$$2x + x = 3x$$
$$\therefore 3x^2 + 2x - 2x^2 + x = x^2 + 3x$$

***Example (3)***   Simplify $4y + 2x + 2y^2 - 3x - 5y$.

***Answer***   $2y^2 - x - y$

***Example (4)***   Simplify $4ab + 3ab - 2ba + 3c$.

***Answer***   $5ab + 3c$   (Note that $ab = ba$.)

## Practice Exercise No. 9

Simplify the following.

(1)  $3a + 5a + 2a$

(2)  $4x + 3y + 2x - y$

(3)  $7z + 2x - 3y + z$

(4)  $3a - 2b - 2a + b$

(5)  $6p - 5q - 7p + 6q$

(6)  $8x - 2z + 16x - 4z$

(7)  $16x^2 - x + x^2 - 5x$

(8)  $9x^3 + 2x^2 - x^3 + 3x^3$

(9)  $16y - 3y^2 + 2y^2 - 8y$

(10)  $7xy - 6yx + 10xy$

(11)  $16ab - 6cd - 3ba + 2dc$

(12)  $14x^2 - 2xy + 3yx + 2x^2$

(13)  $19x - 3xy + 2yx - 21x$

(14)  $5cd + 5c - 3dc$

(15)  $4c^2 + 5c^2 - 3c + 15c$

(16)  $7mn + 6nm - 7m + 6n$

---

*Self-Assessment Test No. 4*

Remove the brackets in the following expressions.

(1)  $x(x - 1)$

(2)  $a(b + c)$

(3)  $y(y^2 + y - 4)$

(4)  $xy(x - y)$

(5)  $xyz(1 + x)$

(6)  $ab(b - c)$

(7)  $pq(r - s)$

(8)  $x^2y^2(x^3 + y^2)$

(9)  $(x - 2)(x + 3)$

---

## 5.4  REMOVAL OF BRACKETS IN AN ALGEBRAIC EXPRESSION

If an expression in brackets is multiplied by a number or letter, each term in the brackets must be multiplied by that number or letter when the brackets are removed.

*Example (1)*  $3(x + y) = 3x + 3y$

*Example (2)*  $5(x^2 + x - 2) = 5x^2 + 5x - 10$

*Example (3)*  $a(b + c - d) = ab + ac - ad$

*Example (4)*  $xy(x + y + z) = x^2y + xy^2 + xyz$

*Example (5)*  $-2(x - y) = -2x + 2y$

## Practice Exercise No. 10

Remove the brackets in the following.

(1)  $3(a + 2)$

(2)  $4(x - y)$

(3)  $6(a + b - c)$

(4)  $9(x^2 - y^2)$

(5)  $a(b - 6)$

(6)  $c(d - 4)$

(7)  $15(x + y - z)$

(8)  $x(x - 3)$

(9)  $xy(z - 2)$

(10)  $ab(c + d)$

(11)  $17(x^2 - x + 3)$

(12)  $x(x^3 - 3x^2 + 2x - 1)$

## 5.5   MISCELLANEOUS EXERCISES

**(1)**   Evaluate the following by using the laws of exponents.

(a)   $\dfrac{6^6 \times 6^2}{6^5 \times 3}$

(b)   $\left(\dfrac{10^5}{10^4}\right)^3$

(c)   $2x^3 \times 3x^2$

**(2)**   Evaluate the following by using the laws of exponents.

(a)   $\dfrac{10^4 \times 10^3}{5 \times 10^6}$

(b)   $\left(\dfrac{5^4}{5^3}\right)^3$

(c)   $8x^6 \div 2x^2$

**(3)**   Evaluate $\dfrac{3^3 \times 3^{-2}}{(3^2)^3 \times 3^{-4}}$ using exponents throughout.

**(4)**   Simplify $\left(\dfrac{x^9 y^6}{y^3}\right)^{1/3}$.

**(5)**   Subtract $3x^2 - 2x + 5$ from $6x^2 + x - 3$.

Simplify the following expressions.

**(6)**   $2p(p - 4)$

**(7)**   $ax(x + 1)$

**(8)**   $2y(y - 2)$

**(9)**   $2(x - 1) + 3(5 - x)$

**(10)**   $3(3x - 2)$

**(11)**   Subtract $5x^2 - 5$ from $7x^2 + 3x - 4$.

Simplify the following expressions.

**(12)**   $4x(1 + 3x)$

**(13)**   $A(A + 4)$

**(14)**   $4a - 6b - 3(a - 2b)$

**(15)**   $a(a + b) - (a^2 - b^2)$

# 6

# Formulas and Equations

---

**Self-Assessment Test No. 1**

| | | |
|---|---|---|
| **(1)** | If $a = 2, b = 3, c = 7$ | Evaluate $2a - 3b + c$ |
| **(2)** | If $x = -1, y = 1, z = -2$ | Evaluate $x^2 + y^2 - z$ |
| **(3)** | If $x = 3, y = 4, z = 0$ | Evaluate $xz + yz + y + x$ |
| **(4)** | If $a = 3, b = 2, c = 1$ | Evaluate $a^2 - b^2 + c$ |
| **(5)** | If $p = -3, q = -2, r = -1$ | Evaluate $p - q - r^2$ |
| **(6)** | If $a = -2, b = 3, c = 1$ | Evaluate $a^2 + b^2 + c^2$ |

---

## 6.1   FORMULA SUBSTITUTION

A formula may be given in algebraic terms; for example,

$$E = IR$$

where

$$E = \text{e.m.f. in volts}$$
$$I = \text{current in amperes}$$
$$R = \text{resistance in ohms}$$

If two values are given, one for current $I$ and one for resistance $R$, then these values substituted in the formula will enable the e.m.f. to be found. For example, if we know that $E = IR$, and that $I = 0.3$ A and $R = 500$ $\Omega$, then we can calculate $E$.

$$E = 0.3 \times 500$$
$$= 150 \text{ V}$$

**84**

If $I = 0.5$ A and $R = 25\ \Omega$, then

$$E = 0.5 \times 25$$
$$= 12.5\ \text{V}$$

## Practice Exercise No. 1

If $a = 3$, $b = 2$, $c = 1$, $d = -1$, and $e = 0$, evaluate the following.

(1) $a + b + c + d + e$     (2) $a + c - 2d$       (3) $2a + 2b - c$

(4) $ab + cd + be$        (5) $a^2 + b^2 + cd$     (6) $\dfrac{ab^2 + cd}{2}$

(7) $\dfrac{2}{3}(d^2 + 2a + b)$

If $x = 1$, $y = -3$, and $z = -2$, evaluate the following.

(8) $x^2 - y^2$         (9) $xy + yz$        (10) $z - xy$

(11) $xyz + y^2$

(12) If $a = bcd$, calculate $a$ when $b = 2$, $c = 3$, and $d = 4$.

(13) If $p = \dfrac{qr}{s}$, calculate $p$ when $q = 5$, $r = 100$, and $s = 4$.

(14) If $P = \dfrac{IW}{T}$, calculate $P$ when $I = 17$, $W = 24$, and $T = 6$.

(15) If $z = \dfrac{abc}{pt}$, calculate $z$ when $a = 40$, $b = 50$, $c = 1$, $p = 10$, and $t = 400$.

## 6.2 PRACTICAL APPLICATIONS
### EXERCISE NO. 1

*(1) The formula connecting rev/min and cutting speed is $N = \dfrac{1000S}{\pi D}$ (see
Fig. 6–1).
Calculate $N$ when
  (a)   $S = 50$ m/min, $D = 54$ mm.
  (b)   $S = 70$ m/min, $D = 35.5$ mm.
  (c)   $S = 40$ m/min, $D = 98$ mm.

(2) The formula for conversion of degrees Fahrenheit into degrees Celsius
is $°C = \dfrac{5}{9}(°F - 32)$. Calculate C when
  (a)   $F = 180°$
  (b)   $F = 36°$

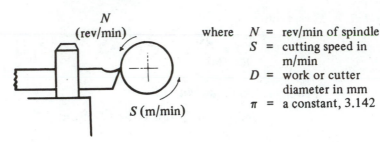

where  $N$ = rev/min of spindle
$S$ = cutting speed in m/min
$D$ = work or cutter diameter in mm
$\pi$ = a constant, 3.142

**Fig. 6–1**

(c)  F = 130°
(d)  F =  77°

(3)  Horsepower may be calculated using the formula $HP = \dfrac{I^2 R}{746}$, where

$$I = \text{current in amperes}$$
$$R = \text{resistance in ohms}$$

Calculate the $HP$ when
(a)  $I$ =   3 A,  $R$ = 1000 $\Omega$.
(b)  $I$ =   2 A,  $R$ = 2000 $\Omega$.
(c)  $I$ =  12 A,  $R$ =  400 $\Omega$.
(d)  $I$ = 1.5 A,  $R$ = 1000 $\Omega$.

(4)  The volume of a cone is given by the formula $V = \dfrac{1}{3}\pi r^2 h$, where

$$r = \text{base radius}$$
$$h = \text{perpendicular height}$$
$$\pi = \text{a constant, 3.142}$$

Calculate $V$ when
(a)  $r$ =   3 ft,  $h$ =   9 ft   ($V$ in ft³).
*(b)  $r$ =   5 cm,  $h$ =   8 cm  ($V$ in cm³).
(c)  $r$ =  15 in,  $h$ =  25 in  ($V$ in in.³).
*(d)  $r$ = 43 mm,  $h$ = 98 mm  ($V$ in mm³).

(5)  When tapping a hole (putting a thread inside), the hole must first be drilled before the thread can be cut (see Fig. 6–2). The tapping drill sizes for Whitworth threads can be calculated using the formula $T = D - 1.1328P$ (see Fig. 6–3), where

$$D = \text{diameter of thread in inches}$$
$$P = \text{pitch of thread in inches}$$
$$T = \text{tapping drill diameter in inches}$$

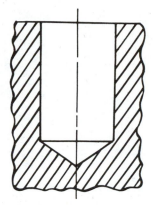

**Fig. 6–2**

Calculate $T$ when $D = \dfrac{3''}{8}$ and $P = 0.0625''$. Answer correct to 3 decimal places.

*(6)   A special rivet is proportioned as shown in Fig. 6–4. Sketch the rivet and dimensions with actual lengths if:

   (a)   $D = 9.5$ mm.

   (b)   $D = 34$ mm.

   (c)   $D = \dfrac{5''}{16}.$

   (d)   $D = 2.4$ cm.

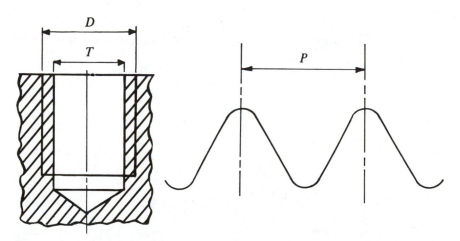

**Fig. 6–3**

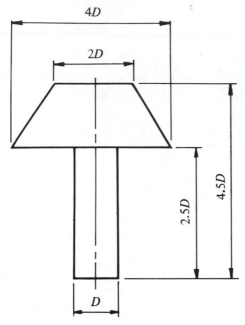

**Fig. 6–4**

## 6.3 FORMULATION OF EQUATIONS (OPTIONAL)

An extruded section is shown in Fig. 6–5, all dimensions of the cross-section being related to the thickness $t$. A range of sizes is to be produced with various values of $t$.

**(1)** Work out a formula for the cross-sectional area of the extrusion in terms of $t$.

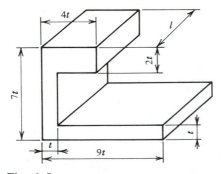

**Fig. 6–5**

**\*(2)** Calculate the cross-sectional area if $t = 20$ mm.

**(3)** If the length of the extrusion is given as $l$, work out a formula for the volume in terms of $t$ and $l$.

**(4)** If $t = 2''$ and $l = 30''$, calculate the volume of the extruded part.

Let's work out each of these problems.

**(1)** Cross-sectional area $A$ = area of $x$ + area of $y$ + area of $z$ (Fig. 6–6).

$$\text{Area } x = 4t \times 2t = 8t^2$$
$$\text{Area } y = (7t - 2t - t) \times t$$
$$= 4t \times t$$
$$= 4t^2$$
$$\text{Area } z = 9t \times t = 9t^2$$

Thus total cross-sectional area $A = 8t^2 + 4t^2 + 9t^2$.

*Answer* $A = 21t^2$

**\*(2)** If $t = 20$ mm,

$$A = 21t^2$$
$$A = 21 \times (20)^2$$
$$= 21 \times 400$$
$$= 8400 \text{ mm}^2$$

*Answer* Cross-sectional area $= 8400$ mm$^2$

**(3)** Volume = cross-sectional area × length. Thus,

$$V = 21t^2 \times l$$
$$V = 21t^2 l$$

*Answer* Volume $= 21t^2 l$

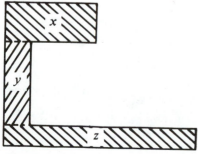

**Fig. 6–6**

**(4)** If $t = 2''$, $l = 30''$, and $V = 21t^2l$,

$$V = 21 \times (2)^2 \times 30$$
$$= 21 \times 4 \times 30$$
$$= 2520 \text{ in.}^3$$

*Answer* Volume $= 2520$ in.$^3$

## 6.4 PRACTICAL APPLICATIONS
## EXERCISE NO. 2

**(1)** Figure 6–7 shows the shape of a component made in a range of sizes.
    **(a)** Obtain an equation, in terms of $x$, for the perimeter $P$ of the shape.
    **(b)** Obtain an equation, in terms of $x$, for the area $A$ of the shape.
    **(c)** Find $P$ and $A$ if $x = 2''$.
    **\*(d)** Find $P$ and $A$ if $x = 30$ mm.
    **(e)** Find $P$ and $A$ if $x = 10''$.
    **\*(f)** Find $P$ and $A$ if $x = 25.9$ mm.
    **(g)** Find $P$ and $A$ if $x = 2.68''$.

**(2)** Figure 6–8 shows the shape of an open box made in a range of sizes.
    **(a)** Obtain an equation, in terms of $h$, $l$, and $w$, for the surface area $A$ of the box.
    **(b)** Obtain an equation, in terms of $h$, $l$, and $w$, for the volume $V$ of the box.
    **(c)** If $l = 3.5''$, $w = 2''$, and $h = 1.5''$, calculate $A$ and $V$.

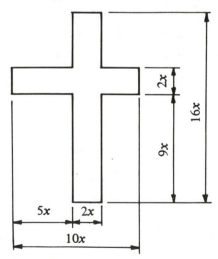

**Fig. 6–7**

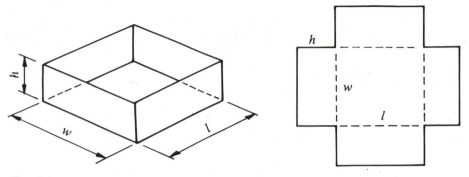

**Fig. 6–8**

*(d)*  If $l$ = 120 mm, $w$ = 115 mm, and $h$ = 50 mm, calculate $A$ and $V$.

---

**Note:**  *If you need help with questions involving volume, refer to problem 3 of Section 6.3 above or to Chapter 11, Section 11.1.*

---

(3)  You are to make a set of fully enclosed cylindrical cans, as shown in Fig. 6–9.

    **(a)**  Work out an expression for the surface area $A$ of the cans in terms of $r$ and $h$. (Area of a circle = $\pi r^2$; circumference = $2\pi r$.)

    **(b)**  Work out an expression for the volume $V$ of the cans in terms of $r$ and $h$.

    *(c)*  If $h$ = 12 cm and $r$ = 5 cm, calculate $A$ and $V$.

    **(d)**  If $h$ = 11.5" and $r$ = 3", calculate $A$ and $V$.

(4)  An end view of a house is shown in Fig. 6–10.

    **(a)**  Calculate the area of brickwork $A$ in terms of $a$, $b$, $h$, and $l$.

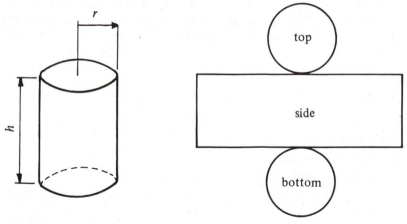

**Fig. 6–9**

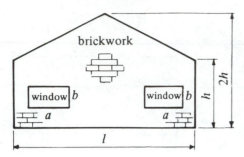

**Fig. 6–10**

*(b)* If $h = 3.3$ m, $a = 1.5$ m, $b = 2.3$ m, and $l = 5.5$ m, calculate $A$.

---

*Self-Assessment Test No. 2*

Solve the following equations.

**(1)** $x + 3 = 7$   **(2)** $2x - 4 = 10$   **(3)** $3x + 2 = 11$

**(4)** $\dfrac{x - 5}{2} = 14$   **(5)** $\dfrac{x}{3} = 12$   **(6)** $\dfrac{2a - 3}{2} = 6$

**(7)** $\dfrac{2(a + 3)}{5} = 10$   **(8)** $\dfrac{x}{2} - 3 = 12$   **(9)** $\dfrac{z}{0.4} = 1.2$

**(10)** $\dfrac{2}{3}(x + 3) = 8$

---

## 6.5   SOLUTION OF EQUATIONS: PART 1

When solving an equation (e.g., $x + 1 = 7$), the value of the unknown ($x$) has to be found. The most important fact to remember when solving equations is that "what you do to one side of an equation, you must also do to the other." Here are some guidelines.

**(1)** Equal numbers may be added to each side.

**(2)** Equal numbers may be subtracted from each side.

**(3)** Each side may be multiplied by equal numbers.

**(4)** Each side may be divided by equal numbers.

*Example (1)*   Solve the equation $x - 3 = 7$.

Add 3 to both sides.

$$x - \cancel{3} + \cancel{3} = 7 + 3$$

***Answer***   $x = 10$

***Check***   $10 - 3 = 7$

*Example (2)* Solve the equation $x + 1 = 7$.
Subtract 1 from both sides.
$$x + \cancel{1} - \cancel{1} = 7 - 1$$
**Answer** $x = 6$
**Check** $6 + 1 = 7$

*Example (3)* Solve the equation $3x = 9$.
Divide both sides by 3.
$$\frac{\cancel{3}x}{\cancel{3}} = \frac{9}{3}$$
**Answer** $x = 3$
**Check** $3 \times 3 = 9$

*Example (4)* Solve the equation $\dfrac{x}{4} = 6$.
Multiply both sides by 4.
$$\frac{x}{\cancel{4}} \times \cancel{4} = 6 \times 4$$
**Answer** $x = 24$
**Check** $\dfrac{24}{4} = 6$

*Example (5)* Solve the equation $3x + 2 = 11$.
Subtract 2 from both sides.
$$3x + \cancel{2} - \cancel{2} = 11 - 2$$
$$3x = 9$$
Divide both sides by 3.
$$\frac{\cancel{3}x}{\cancel{3}} = \frac{9}{3}$$
**Answer** $x = 3$
**Check** $3 \times 3 + 2 = 11$

## Practice Exercise No. 2

Solve the following equations.

(1)  $x + 2 = 18$

(2)  $x - 3 = 10$

(3)  $x + 6 = 4$

(4)  $p + 7 = 4$

(5)  $t - 6 = 14$

(6)  $7 + s = 8$

(7)  $8 - t = 3$

(8)  $x - 14 = 0$

(9)  $3p = 15$

(10)  $4x = 132$

(11)  $9y = 11.7$

(12)  $\dfrac{s}{2} = 8$

(13)  $\dfrac{t}{8} = \dfrac{1}{5}$

(14)  $\dfrac{Q}{8} = 1.2$

(15)  $2s + 3 = 11$

(16)  $4x - 8 = 6.4$

(17)  $\dfrac{x}{0.4} = 10$

(18)  $4y + 2 = 7.2$

## 6.6 SOLUTION OF EQUATIONS: PART 2

In an equation involving several operations, the best approach is to "unravel" the operations. In other words, as you would "unravel" a knot in the reverse order of tying it, you should solve an equation in this way.

**(1)** Figure out the order in which the operations would be done if you knew the value to substitute.

**(2)** Reverse those operations ($\times$ for $\div$ and $+$ for $-$, etc.) in the opposite order.

For Example (1) which follows, the order of operations for $\frac{1}{2}(3x + 2)$ is: multiply by 3 (if you knew what $x$ was), add 2, and divide by 2.

We solve this equation by doing the following operations to each side:

| | |
|---|---|
| Multiply by 2 | (reverse of divide by 2) |
| Subtract 2 | (reverse of add 2) |
| Divide by 3 | (reverse of multiply by 3) |

***Example (1)*** Solve the equation $\frac{1}{2}(3x + 2) = 7$.

Multiply both sides by 2.

$$\cancel{2} \times \frac{1}{\cancel{2}}(3x + 2) = 7 \times 2$$

$3x + 2 = 14$

Subtract 2 from both sides.

$3x + \cancel{2} - \cancel{2} = 14 - 2$

$3x = 12$

Divide both sides by 3.

$$\frac{\cancel{3}x}{\cancel{3}} = \frac{12}{3}$$

***Answer*** $x = 4$

***Check*** $\frac{1}{2}(3 \times 4 + 2) = 7$

***Example (2)*** Solve the equation $x^2 = 9$.

Take the square root of both sides.

$x = 3$ or $x = -3$

| ***Answer*** $x = 3$ | ***Answer*** $x = -3$ |
|---|---|
| ***Check*** $3^2 = 9$ | ***Check*** $(-3)^2 = 9$ |

*Example (3)*    Solve the equation $2x - 3 = 3x - 4$.
In this case we must get the letter terms together.
Subtract $2x$ from both sides.
$2\cancel{x} - 2\cancel{x} - 3 = 3x - 2x - 4$
$\qquad\qquad - 3 = x - 4$
Add 4 to both sides.
$-3 + 4 = x - \cancel{4} + \cancel{4}$
$\qquad 1 = x$

*Answer*  $x = 1$

*Check*   LHS $= 2 \times 1 - 3 = -1$
$\qquad\quad$ RHS $= 3 \times 1 - 4 = -1$
(Left- and right-hand sides are equal.)

*Example (4)*

Solve the equation $\dfrac{5x - 1}{2} = 7x + 4$.

Multiply both sides by 2.
$\dfrac{\cancel{2}(5x - 1)}{\cancel{2}} = 2(7x + 4)$
$5x - 1 = 14x + 8$
Subtract $5x$ from both sides.
$\cancel{5}x - \cancel{5}x - 1 = 14x - 5x + 8$
$\quad - 1 = 9x + 8$
Subtract 8 from both sides.
$-1 - 8 = 9x + \cancel{8} - \cancel{8}$
$\quad - 9 = 9x$
Divide both sides by 9.
$$\dfrac{\overset{1}{-\cancel{9}}}{\underset{1}{\cancel{9}}} = \dfrac{\overset{1}{\cancel{9}}x}{\underset{1}{\cancel{9}}}$$

*Answer*  $x = -1$

*Check*  LHS $= \dfrac{5 \times (-1) - 1}{2} = -3$
$\qquad\quad$ RHS $= 7 \times (-1) + 4 = -3$

---

**Note:**  *When you have arrived at an answer, always substitute the value obtained back into the original equation to check if it is correct.*

### Practice Exercise No. 3

Solve the following equations.

**(1)**  $\dfrac{2(x + 1)}{3} = 4$        **(2)**  $4x^2 = 36$        **(3)**  $\dfrac{7y}{3} - \dfrac{2y}{6} = 4$

**(4)**  $\dfrac{2}{3}(y + 1) = \dfrac{4}{3}$        **(5)**  $\dfrac{t}{2} - \dfrac{3}{2} = 7$        **(6)**  $3x - 5 = 2x + 3$

**(7)**  $3y + 4 = y - 8$        **(8)**  $y - 5 = \dfrac{5 - y}{3}$        **(9)**  $\dfrac{2y - 3}{2} = 3y - 10$

## 6.7   PRACTICAL APPLICATIONS EXERCISE NO. 3

**\*(1)**   The formula for the area of a square $A$, of side $a$, is $A = a^2$. Calculate $a$ when $A = 144$ cm².

**\*(2)**   The perimeter $P$ of a rectangle is given by the formula $P = 2a + 2b$ where $a$ and $b$ are the lengths of its sides.
   **(a)**   Calculate $a$ if $P = 7.88$ m and $b = 1.69$ m.
   **(b)**   Calculate $b$ if $P = 64$ mm and $a = 18$ mm.

**(3)**   The area $A$ of a rectangle is given by the formula $A = lb$ where $l =$ length and $b =$ width.
   **\*(a)**   If a rectangle has an area of 960 mm² and length of 120 mm, calculate $b$.
   **(b)**   If a rectangle has an area of 1400 in.² and width of 20 in., calculate $l$.

**(4)**   Two bars are measured and are found to have lengths $m$ and $n$. The average of these two lengths, $x$, is given by the formula
$$x = \frac{m + n}{2}.$$
   **(a)**   If $x$ is 18 in. and $m$ is 7 in., calculate $n$.
   **\*(b)**   If $x$ is 235 mm and $n$ is 146 mm, calculate $m$.
   **(c)**   If $x$ is 43.6 ft and $m$ is 33.7 ft, calculate $n$.

**(5)**   The cost $\$C$ of $x$ number of drills costing 64¢ each is given by the formula $C = \dfrac{\$64x}{100}$. Use the formula to find how many drills can be purchased for **(a)** $128, **(b)** $32, **(c)** $84.48, **(d)** $40.32, and **(e)** $49.92.

**(6)**   The cost of producing a certain component consists of setting up the machine, materials, and production costs. If the setting up cost *per batch* is $3.80 and material/production costs total 18¢ per component, obtain a formula for the total cost $\$C$ of a batch of $x$ components. If the total cost for a certain batch was $309.80, how many components were made?

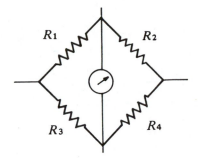

**Fig. 6–11**

(7) The formula connecting power (*P*) in watts, current (*I*) in amperes, and e.m.f. (*E*) in volts is $P = EI$.

   (**a**) Calculate *I* when $P = 500$ W and $E = 240$ V.

   (**b**) Calculate *E* when $P = 300$ W and $I = 12.5$ A.

\*(8) The formula for the conversion of degrees Celsius into degrees Fahrenheit is given as $°F = \frac{9}{5} °C + 32$. Use the formula to find the value of F when (**a**) C = 35°, (**b**) C = 80°, (**c**) C = 75°, and (**d**) C = 24°.

(9) The resistances in an electrical Wheatstone Bridge circuit, shown in Fig. 6–11, are connected by the formula

$$\frac{R_1}{R_2} = \frac{R_3}{R_4}$$

   If $R_1 = 30$ Ω, $R_3 = 40$ Ω, and $R_4 = 60$ Ω, calculate $R_2$.

(10) If *x* and *y* are two resistances connected in parallel, the total resistance *R* is given by the formula $\frac{1}{R} = \frac{1}{x} + \frac{1}{y}$ (see Fig. 6–12).

   (**a**) Calculate the value of *x* when $R = 4$ Ω and $y = 8$ Ω.

   (**b**) Calculate the value of *y* when $R = 16$ Ω and $x = 64$ Ω.

(11) The volume of a cylindrical can (Fig. 6–13) is given by the formula $V = \pi r^2 h$.

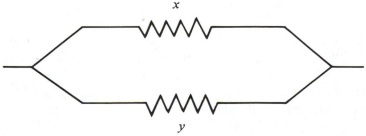

**Fig. 6–12**

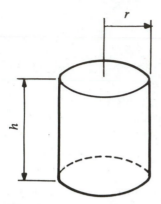

**Fig. 6–13**

    **(a)**   Calculate the radius of a can that is to hold 2000 cm³ of liquid if the height is to be 20 cm.

    **(b)**   Calculate the height of a tank that is to hold 4.94 yd³ of liquid if the radius is 1.1 yd.

---

*Self-Assessment Test No. 3*

Transpose the following formulas, making the symbol indicated the subject.

**(1)** $x = a + 4$ for $a$      **(2)** $x + a = 5$ for $a$      **(3)** $pv = 500$ for $v$

**(4)** $\dfrac{p}{5} = x$ for $p$      **(5)** $a + 3 = bc$ for $c$      **(6)** $5 = 2\pi r$ for $r$

**(7)** $x^2y = 4$ for $y$      **(8)** $x^2y = 4$ for $x$      **(9)** $\dfrac{x - y}{3} = t$ for $x$

**(10)** $\dfrac{x - y}{3} = t$ for $y$

---

## 6.8 TRANSPOSITION OF FORMULAS

**Transposition** means the rearrangement of the symbols in a formula or equation in order to make a different symbol the subject of the formula. For example, the formula $v = u + at$ may be transposed to make $u$ the subject: $u = v - at$.

    Transposition is carried out by adding, subtracting, multiplying, or dividing both sides of an equation by the same quantity (similar to the last section).

*Example (1)*    Transpose the formula $a = b + x$ to make $x$ the subject.

    Subtract $b$ from both sides.

    $a - b = \not b + x - \not b$

    **Answer**  $x = a - b$

*Example (2)*    Transpose the formula $A = \dfrac{B^2C}{3Z}$ to make $B$ the subject.

Multiply both sides by $3Z$.

$$3ZA = \frac{B^2C}{\cancel{3Z}} \cdot \cancel{3Z}$$

Divide both sides by $C$.

$$\frac{3ZA}{C} = \frac{B^2\cancel{C}}{\cancel{C}}$$

Take the square root of both sides.

$$\sqrt{\frac{3ZA}{C}} = \sqrt{B^2}$$

*Answer*    $B = \sqrt{\dfrac{3ZA}{C}}$

## Practice Exercise No. 4

Transpose the following formulas, making the symbol indicated the subject.

**(1)**  $a = b + c$ for $c$     **(2)**  $y = bx$ for $x$     **(3)**  $s = \dfrac{x}{4}$ for $x$

**(4)**  $7x - y = 4$ for $y$     **(5)**  $t = \dfrac{4}{7s}$ for $s$     **(6)**  $N = \dfrac{SD}{PR}$ for $D$

**(7)**  $\dfrac{xy}{z} = 4$ for $y$     **(8)**  $\dfrac{4 - t}{2} = 3p$ for $t$     **(9)**  $\dfrac{7 + p}{t} = z$ for $p$

**(10)**  $\dfrac{8 + zp}{t} = y$ for $p$

## 6.9  PRACTICAL APPLICATIONS EXERCISE NO. 4

**(1)**  The formula connecting rev/min and cutting speed is $N = \dfrac{12S}{\pi D}$, where

$N$ = rev/min of the spindle

$S$ = cutting speed in ft/min

$D$ = work or cutter diameter in inches

$\pi$ = a constant, 3.142

Transpose the formula to make
**(a)**  $S$ the subject.
**(b)**  $D$ the subject.

(c)   Calculate $N$ when $S = 150$ ft/min and $D = \dfrac{3''}{4}$.

(d)   Calculate the rev/min for a 2″ bar to cut at 80 ft/min.

*(2)   The metric version of the formula in Question 1 is $N = \dfrac{1000S}{\pi D}$, where

$$N = \text{rev/min of the spindle}$$
$$S = \text{cutting speed in m/min}$$
$$D = \text{work or cutter diameter in mm}$$
$$\pi = \text{a constant, } 3.142$$

(a)   Transpose the formula to make $S$ the subject.
(b)   Calculate $N$ when $S = 30$ m/min and $D = 18$ mm.
(c)   Calculate $S$ when $N = 600$ rev/min and $D = 35$ mm.

*(3)   The formula for conversion of degrees Fahrenheit into degrees Celsius is $°C = \dfrac{5}{9}(°F - 32)$.

(a)   Transpose the formula to make F the subject.
(b)   Convert 98°C into degrees Fahrenheit.
(c)   Convert 45°C into degrees Fahrenheit.
(d)   Convert 0°C into degrees Fahrenheit.

(4)   Horsepower may be calculated using the formula $HP = \dfrac{I^2 R}{746}$, where

$$I = \text{current in amperes}$$
$$R = \text{resistance in ohms}$$

Transpose the formula to make
(a)   $R$ the subject.
(b)   $I$ the subject.
(c)   Calculate $HP$ if $R = 3000\ \Omega$ and $I = 3$ A.
(d)   Calculate $I$ if $HP = 4.5$ and $R = 1000\ \Omega$.
(e)   Calculate $R$ if $HP = 10$ and $I = 5$ A.

(5)   A worker on piecework gets a basic wage of $15 per day plus 45¢ per article made. Wages per day can be calculated using the formula

$$W = 15 + \frac{45x}{100}$$

where
$$W = \text{wages in dollars}$$
$$x = \text{number of components produced}$$

(a)   Transpose the formula to make $x$ the subject.

**(b)** If 87 components are produced in one day, what will be the wage?

**(c)** If the worker wishes to earn $80 in one day, how many articles must be made?

**(d)** If the worker wishes to earn $95.50 in one day, how many articles must be made?

**(6)** The formula connecting power (in watts), e.m.f. (in volts), and current (in amperes) is $P = EI$.

$$(\text{watts}) = (\text{volts}) \cdot (\text{amperes})$$

**(a)** Transpose the formula to make $I$ the subject.

**(b)** You are required to fit plugs to the following electrical appliances.

| | | |
|---|---|---|
| Iron | rated | 750 W |
| Hairdryer | rated | 525 W |
| Heater | rated | 2 kW |
| Refrigerator | rated | 120 W |
| Television | rated | 300 W |

You have a choice of 5 A, 10 A, 15 A, and 20 A fuses. Calculate for each appliance the current drawn when connected to a 120 V supply and also state the size of fuse required for each appliance.

## 6.10  MISCELLANEOUS EXERCISES

**(1)** If $3.5x = 10.5$, what is the value of $x$?

**(2)** Solve the following equations.
  **(a)** $2(y - 5) = 10 - 6y$
  **(b)** $5 - x = 2(x - 2)$

**(3)** Transpose the formula $P = I^2R$ to make $I$ the subject, and determine the value of $I$ when $P = 810$ and $R = 40$.

**(4)** Solve the following equations.
  **(a)** $p - 4 = 4(p - 2)$
  **(b)** $\dfrac{x + 3}{2} = 10$

**(5)** Transpose the formula $p = \dfrac{ksv^2}{a}$ to make $v$ the subject, and determine the value of $v$ when $p = 70$, $a = 7$, $k = 0.01$, and $s = 9000$.

**(6)** Make $E$ the subject of the formula

$$R + r = \frac{E - e}{I}$$

and then calculate its value when $R = 20\ \Omega$, $r = 2\ \Omega$, $e = 60$ V, and $I = 6$ A.

(7) Rearrange the formula $V = \dfrac{\pi D^2 L}{4}$ to make $L$ the new subject.

(8) If $\dfrac{VA}{1000} = K$, what will be the value of $V$ when $K = 4$ and $A = 20$?

(9) For the following solutions, give your answer correct to two significant digits.
  (a)  $V = \pi r^2 h$. Make $r$ the subject of the formula.
  (b)

$$\text{If } S = \frac{1}{2}at^2, \text{ find } S \text{ when } a = 12.4 \text{ and } t = 2.34.$$

(10) Express, in the simplest terms of $a$ and $b$, as illustrated in Fig. 6–14,
  (a)  Perimeter of the figure
  (b)  Area of the figure

(11) Solve the following equations.
  (a)  $3x = 24$
  (b)  $3x + 6 = 24$
  (c)  $6x - 3 = 24$
  (d)  $\dfrac{2x}{5} = \dfrac{3}{10}$

(12) A strip of metal with a length of 20 units is bent to form a rectangle. The ends are buttwelded without overlap.
  (a)  Denoting the length of the rectangle by $x$, obtain an expression for the width.
  (b)  Prove that the area $A$ of the rectangle can be obtained from the formula

$$A = 10x - x^2$$

  (c)  Find the dimensions of the rectangle when $A = 24$.

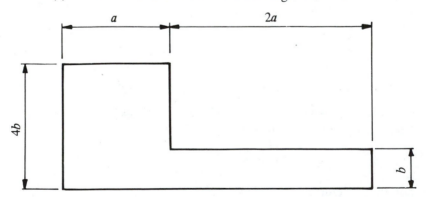

Fig. 6–14

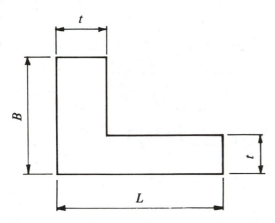

**Fig. 6–15**

**(13)** Make $g$ the subject of the formula

$$W(h + x) = mgh$$

**(14)** Make $L$ the subject of the formula

$$d = \frac{5WL^3}{384EI}$$

**(15)** Solve the equation $5(2x - 7) - 4(3x + 1) + 45 = 0$.

**\*(16)** The perimeter of triangle $ABC$ is 1.2 m. Side $AB$ is three-fifths of side $BC$ and side $AC$ is 100 mm shorter than $BC$. Calculate the lengths of the sides of the triangle.

**(17)** Transpose the formula $I = \dfrac{nE}{R + nr}$ to make $E$ the subject.

**(18)** Find an expression for $V$ from the gas formula $pV = r(1 + at)$.

**(19)** The surface area $A$ of an open topped cylindrical vessel is given by $A = \pi R^2 + 2\pi RH$. Transpose the expression to give $H$ in terms of the other symbols.

**(20)** If $V = \dfrac{1}{3}\pi r^2 l$, express $r$ as the subject of the equation.

**(21)** Solve the following equation for $p$.

$$0.8(8p - 1.3) = 1.76 + 5p$$

**(22)** Solve the equation $5(y - 3) + 43 = 3(4 - y)$.

**(23)** Deduce a formula for the area $A$ of the angle section shown in Fig. 6–15.

# 7

# Ratio and Proportion

**Self-Assessment Test**

Write each of the following ratios in its simplest form.

**(1)** 7 to 14    **(2)** 33 to 11    **(3)** 75 to 15    **(4)** $\frac{1}{4}$ to 4    **(5)** 144 to 72

**(6)** Divide $700 in the ratio of 1:3:6.

**(7)** Divide $1400 in the ratio of 3:6:11.

**(8)** Divide 360° in the ratio of 1:3:8.

**(9)** A worker makes 14 components in 4 hours; how many will be made in 18 hours?

**(10)** A gear $A$ has 100 teeth and makes 84 rev/min. It is in mesh with a gear $B$ which has 75 teeth, Calculate the rev/min of $B$.

## 7.1 RATIO

A **ratio** is the relation between two like values or numbers. If a car is 5 m long and a model of it is 1 m long, then the length of the model is 1/5th of the length of the car. All dimensions of the car are reduced in the ratio 1 to 5 when making the model.

A ratio may be written as a fraction. For example,

$$\frac{1}{5} \quad \text{or} \quad 1:5 \quad \text{or} \quad 1 \text{ to } 5$$

Since a ratio can be regarded as a fraction, multiplying or dividing both terms of a ratio by the same number does not change the value of the ratio.

For example,

| | |
|---|---|
| 2:3 = 12:18 | Multiplying both sides by 6 |
| 14:7 = 2:1 | Dividing both sides by 7 |
| 27 to 3 = 9 to 1 | Dividing both sides by 3 |
| $\frac{1}{6}$ to 1 = 1 to 6 | Multiplying both sides by 6 |

> **Note:** *Units of both parts of a ratio should be the same, if they are the same type of measure (money, distance, weight, etc.)*

$$\$1:5¢ = 100¢:5¢ = 20:1$$
$$1 \text{ m}:50 \text{ cm} = 100 \text{ cm}:50 \text{ cm} = 2:1$$

### Practice Exercise No. 1

Write each of the following ratios in its simplest form.

**(1)** 3 to 30      **(2)** 8 to 4      **(3)** $\frac{1}{4}$ to 5

**(4)** $\frac{1}{3}$ to 1      **(5)** 3 to $3\frac{1}{2}$      **(6)** $\frac{1}{4}$ to $\frac{1}{2}$

**\*(7)** 60 m to 20 m      **(8)** \$5 to 50¢      **(9)** 1 lb to 20 oz

**(10)** 10 ft to 50″      **(11)** 1 week to 3 days      **(12)** 30 min to 6 h

**(13)** $4\frac{1}{5}$ to $\frac{2}{5}$      **\*(14)** 3 kg to 99 kg      **(15)** 6 h to 3 days

**(16)** 88 teeth to 22 teeth      **(17)** $6\frac{1}{4}$ to $3\frac{1}{8}$      **\*(18)** 4 km to 30 m

## 7.2 TO SEPARATE A QUANTITY ACCORDING TO A GIVEN RATIO

***Example (1)***   \$50 is to be divided into two portions in the ratio 3:7. How much is in each portion?

***Answer***   Total number of shares or parts = 3 + 7 = 10.

One portion is $\frac{3}{10}$ of the money:

$$\frac{3}{10} \times \$50 = \$15$$

The other portion is $\dfrac{7}{10}$ of the money:

$$\frac{7}{10} \times \$50 = \$35$$

A similar method is used if more than two portions are involved.

***Example (2)*** $46 has to be split among Peter, Paul, and Ben in the ratio of 3:9:11. How much does each receive?

***Answer*** Total number of shares (or parts) = 3 + 9 + 11 = 23.

Peter receives $\dfrac{3}{23}$ of the money:

$$\frac{3}{\cancel{23}_{1}} \times \$ \cancel{46}^{2} = \$6$$

Paul receives $\dfrac{9}{23}$ of the money:

$$\frac{9}{\cancel{23}_{1}} \times \$ \cancel{46}^{2} = \$18$$

Ben receives $\dfrac{11}{23}$ of the money:

$$\frac{11}{\cancel{23}_{1}} \times \$ \cancel{46}^{2} = \$22$$

***Example (3)*** A line is to be divided into 4 sections in the ratio of 1:2:3:6. If the line is 960 mm long, find the length of each part.

***Answer*** Total number of shares (or parts) = 1 + 2 + 3 + 6 = 12.

$$\text{1st section} = 1 \text{ part} = \frac{1}{12} \text{ of } 960 \text{ mm} = \frac{1}{\cancel{12}_{1}} \times \cancel{960}^{80} \text{ mm} = 80 \text{ mm}$$

$$\text{2nd section} = 2 \text{ parts} = \frac{2}{12} \text{ of } 960 \text{ mm} = 160 \text{ mm}$$

$$\text{3rd section} = 3 \text{ parts} = \frac{3}{12} \text{ of } 960 \text{ mm} = 240 \text{ mm}$$

$$\text{4th section} = 6 \text{ parts} = \frac{6}{12} \text{ of } 960 \text{ mm} = 480 \text{ mm}$$

---

**Note:** *The quantities obtained after dividing in the given ratios must total the original quantity to be divided.*

## 7.3 PRACTICAL APPLICATIONS
### EXERCISE NO. 1

(1) Divide $950 in the ratio of 7 to 3.

(2) Divide $4.50 in the ratio of 1:2:7.

(3) Divide $88 in the ratio of 2 to 7 to 13.

*(4) A brass bar is made from copper and zinc and mixed in the ratio of 7:3 by mass. How much of each will be needed to make a bar that has a mass of 37 kg?

(5) A substance is made from 34 parts lead, 26 parts copper, 30 parts iron, and 35 parts water.
  (a) How much water is there in 7 lb of the substance?
  (b) How much substance could be made with 78 lb of copper?

(6) In a factory, the ratio of wages to materials cost is 5:4. What is the wages bill in a week when total expenditure comes to $90 540?

(7) A cutting oil is made from oils $X$ and $Y$ mixed in the ratio of 8:15.
  (a) How much of each is contained in 23 quarts of cutting oil?
  (b) How much of $Y$ should be added to 18 quarts of $X$?

(8) A plumber and an apprentice earn $25.50 bonus in one week. If they share it in the ratio 5:2, how much will each receive?

*(9) A bricklayer mixes mortar in the ratio 1 part cement to 3 parts sand by volume. How much of each will be used to make 1.2 m³ of mortar?

## 7.4 PROPORTION

A **proportion** is a statement that says that two ratios are equal. The proportion sign is :::; or the sign of equality ($=$) may be used. For example,

$$1:3 \ :: \ 9:27 \qquad \text{or} \qquad \frac{1}{3} = \frac{9}{27}$$

## 7.5 DIRECT PROPORTION

Let quantities $X$ and $Y$ be related to each other. If the amount $X$ increases, this may cause $Y$ to increase also. If the ratio $X{:}Y$ remains constant, then the relationship is one of direct proportion. Here are two examples of direct proportion.

(1) For a car traveling at constant speed, the longer the time, the greater the distance traveled.

(2) The larger the quantity purchased, the higher the total cost, assuming that no reduction is allowed on larger quantities.

**Example (1)**   In a day 14 workers assemble 42 units. How many units would 23 workers assemble?

**Method A**   14 workers assemble 42 units.
23 workers assemble $x$ units.

$$42:x = 14:23 \qquad \text{or} \qquad \frac{42}{x} = \frac{14}{23}$$

Multiply both sides by $x$:

$$\frac{42}{\cancel{x}} \cancel{x} = \frac{14x}{23}$$

Multiply both sides by 23:

$$23 \times 42 = \frac{14x}{\cancel{23}} \times \cancel{23}$$

Divide both sides by 14:

$$\frac{23 \times \overset{3}{\cancel{42}}}{\underset{1}{\cancel{14}}} = \frac{\cancel{14}x}{\cancel{14}} \qquad \therefore \ x = 69$$

**Answer**   Number of machines assembled by 23 workers is 69.

**Method B**   14 men assemble 42 units.

$$\therefore \text{1 worker assembles } \frac{42}{14} \text{ units}$$

$$\therefore \text{23 men assemble } \frac{\overset{3}{\cancel{42}}}{\underset{1}{\cancel{14}}} \times 23 \text{ units}$$

$$= 69 \text{ units}$$

**Answer**   Number of machines assembled by 23 workers is 69.

**Example (2)**   A worker makes 85 components in 5 hours, how many will be made in 3 hours?

**Method A**   In 5 hours the worker makes 85 components.
In 3 hours the worker makes $x$ components.

$$5:3 = 85:x \qquad \text{or} \qquad \frac{5}{3} = \frac{85}{x}$$

Multiply both sides by $x$:

$$\frac{5x}{3} = \frac{85}{\cancel{x}} \, \cancel{x}$$

Multiply both sides by 3:

$$\frac{5x}{\cancel{3}} = \cancel{3} = 85 \times 3$$

Divide both sides by 5:

$$\frac{\cancel{5}x}{\cancel{5}} = \frac{\overset{17}{\cancel{85}} \times 3}{\underset{1}{\cancel{5}}} = 51 \qquad \therefore x = 51$$

*Answer*   In 3 hours the worker would make 51 components.

*Method B*   In 5 hours 85 components are made.

Thus, in 1 hour $\dfrac{\overset{17}{\cancel{85}}}{\underset{1}{\cancel{5}}}$ = 17 components are made.

$\therefore$ in 3 hours $17 \times 3 = 51$ components are made.

*Answer*   In 3 hours the worker would make 51 components.


## 7.6   PRACTICAL APPLICATIONS
### EXERCISE NO. 2

(1)   If a man earns $250 for a 45-hour week, how much will he earn in 60 hours?

*(2)   A spring increases in length by 5.5 cm when a pull of 5 newtons (5 N) is applied. What will be the increase when a pull of 9 N is applied?

(3)   A car travels at constant speed. It goes 80 miles in 55 minutes. How far will it travel in 77 minutes?

(4)   A buyer purchases 21 drills for $24. How much would 49 drills cost?

(5)   A metal bar increases in length by 0.006″ when heated through 4°C. How much will the increase be when it is heated through 7.5°C?

(6)   A firm receives $145 for 800 lb of scrap metal. How much would it receive for 1250 lb?

(7)   A company decides to reduce all its prices in an attempt to sell more goods. All items are marked down in the same proportion. If a hand

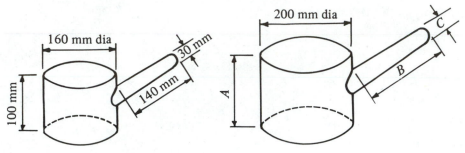

**Fig. 7–1**

drill is marked down from $45 to $39, what would be the new price of a tool set originally selling at $25?

**(8)**   A woman wishes to earn $300.00 per week. She is paid $0.75 for every 10 objects she produces. How many must she make in order to earn the amount required?

**(9)**   A shaft 18″ long is found to have a taper of 0.036″ in 4″. Find the taper in the entire length of shaft.

**\*(10)**   A set of saucepans is to be manufactured. Their sizes are proportional to each other. From the dimensions given in Fig. 7–1, calculate $A$, $B$, and $C$.

**(11)**   A wire 200 ft long has a resistance of 1.14 Ω. What is the resistance of a wire 950 ft long? (Resistance is proportional to length.)

## 7.7   INVERSE PROPORTION

An **inverse proportion** is one in which two quantities are related such that an increase in one causes a corresponding decrease in the other. Here are two examples of inverse proportion.

**(1)**   The more people working, the shorter the time a job takes.

**(2)**   The slower the speed, the longer the time.

Inverse proportion is very important when dealing with gear ratios and pulley/belt drives.

**\*Example (1)**   The two pulleys in Fig. 7–2 are connected by a belt. Pulley $A$ is 20 cm in diameter and runs at 400 rev/min. Pulley $B$ is 35 cm in diameter. Calculate the rev/min of pulley $B$.

Thus will $B$ run slower or faster than $A$?

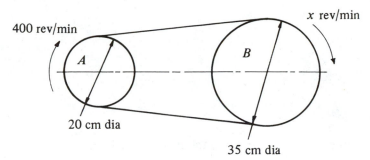

Fig. 7–2

*Answer*  Slower.

$$\frac{\text{rev/min of } B}{\text{rev/min of } A} = \frac{\text{diameter of } A}{\text{diameter of } B} \qquad (\textit{inverse proportion})$$

$$\therefore \frac{x}{400} = \frac{20}{35}$$

Multiply both sides by 400:

$$\cancel{400} \times \frac{x}{\cancel{400}} = \frac{\overset{4}{\cancel{20}}}{\underset{7}{\cancel{35}}} \times 400 = \frac{1600}{7} = 228\frac{4}{7}$$

*Answer*  Pulley $B$ rotates at 229 rev/min (to the nearest rev/min).

*Example (2)*  Two gears are in mesh in Fig. 7–3. Gear $A$ has 48 teeth and rotates at 100 rev/min. Gear $B$ has 72 teeth. At what speed does gear $B$ rotate?

*Answer*  $\dfrac{\text{rev/min of } B}{\text{rev/min of } A} = \dfrac{\text{number of teeth on } A}{\text{number of teeth on } B}$  $\qquad (\textit{inverse proportion})$

$$\therefore \frac{x}{100} = \frac{48}{72}$$

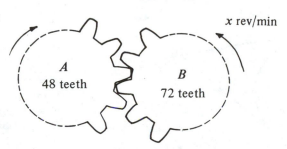

Fig. 7–3

Multiply both sides by 100:

$$\frac{x}{\cancel{100}} \times \cancel{100} = \frac{\overset{2}{\cancel{48}}}{\underset{3}{\cancel{72}}} \times 100 = 66\frac{2}{3}$$

*Answer* Gear B rotates at 67 rev/min (to the nearest rev/min).

## 7.8 PRACTICAL APPLICATIONS
### EXERCISE NO. 3

For Questions 1 to 7, make a neat sketch before solving the problem.

**(1)** A gear having 88 teeth meshes with one having 22 teeth. How many revolutions will the small gear make when the large one revolves once?

**(2)** Gear A has 120 teeth and makes 80 rev/min. It is in mesh with gear B which has 75 teeth. Calculate the number of rev/min made by B.

**(3)** A gear running at 420 rev/min has 55 teeth. It is required to drive another gear at 660 rev/min. How many teeth must the driven gear have?

**\*(4)** A belt drive system is shown in Fig. 7–4. Pulleys B and C are fixed together. If pulley A rotates at 450 rev/min, calculate the speeds of pulleys B, C, and D and state their directions of rotation, that is, clockwise (↻) or counterclockwise (↺).

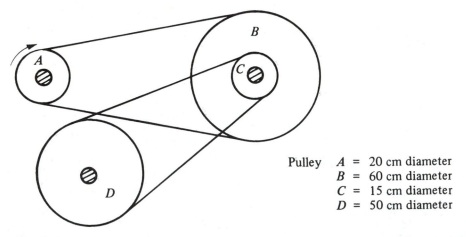

| Pulley | | |
|---|---|---|
| A | = | 20 cm diameter |
| B | = | 60 cm diameter |
| C | = | 15 cm diameter |
| D | = | 50 cm diameter |

**Fig. 7–4**

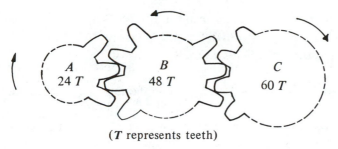

(*T* represents teeth)

**Fig. 7–5**

**(5)** Three gears are in mesh as shown in Fig. 7–5. If *A* makes 480 rev/min, how many rev/min will gears *B* and *C* make?

**(6)** Two gears are to have a speed ratio of 3.5 to 4. If the larger gear has 88 teeth, how many teeth must the smaller one have?

**(7)** In Fig. 7–6, the smaller pulley is 15″ in diameter and the larger pulley 25″ in diameter. How many rev/min will the large pulley make if the smaller pulley rotates at 175 rev/min?

**(8)** The current flowing through a wire decreases as the resistance increases. If the current is 6 A when the resistance is 8 $\Omega$:
    **(a)** What would be the current when the resistance is 10 $\Omega$?
    **(b)** What would be the resistance when the current is 1.5 A?

**(9)** Six workers take 42 days to complete a job. How long would 21 workers have taken?

**(10)** A mixer rotating at 50 rev/min takes 4.5 min to mix a cake. How long would it have taken if it had been rotating at 450 rev/min?

## 7.9 MISCELLANEOUS EXERCISES

**\*(1)** Divide a line 180 mm long into three parts in the ratio of 1:2:3 by calculation.

**(2)** If 24 bushes cost $44.88, what would be the cost of 36 bushes?

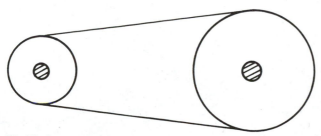

**Fig. 7–6**

(3) An alloy consists of lead and tin in the ratio 7:3. What will be the weight of lead and the weight of tin in 210 ounces of the alloy?

(4) Lathe tools are priced at $33 for 12 tools. What will be the cost of 18 tools?

(5) A certain brass consists of copper and zinc in the ratio 17:8. Calculate the mass of each metal required to make 100 lb of brass.

*(6) Two turns of the screw of a jack raise the load 12.8 mm. Calculate the distance moved by the load for fifteen turns of the screw.

(7) The total amount of money earned by four workers in a week is $1044. If the workers' wages are in the ratios of 9:8:7:5, calculate each worker's wage.

(8) A casting with the mass of 120 lb consists of copper, zinc, and manganese in the ratio of 60:39:1 (by mass). Calculate the mass of each constituent in the casting.

*(9) A 200-mm diameter pulley is fitted to the headstock drive shaft of a lathe and is belt driven from a 100-mm diameter pulley rotating at 1500 rev/min. Calculate the speed of the headstock pulley.

*(10) The speed of a belt conveyor is directly proportional to the diameter of its driving drum. With a drum of diameter 0.5 m the belt speed is 80 m/min.

   (a) Find the diameter of the drum required to give a belt speed of 100 m/min.

   (b) Find the belt speed obtained with a drum of diameter 0.7 m.

(11) The quantity of air delivered by a fan is directly proportional to the fan speed. A mine fan delivers 60 yd³/s of air when its speed is 300 rev/min. Determine the quantity it will deliver when its speed is increased to 400 rev/min.

# PART III

## Applied Geometry

# 8

## Squares, Square Roots, and the Theorem of Pythagoras

> **Calculator Hint**
>
> If you intend to use a calculator to find squares and square roots, you may wish to turn to Section C.3 in Appendix C. With your instructor's permission, you may skip over Sections 8.1 and 8.2 and work only Practice Exercise Nos. 1, 2, and 3 on your calculator.

> *Self-Assessment Test No. 1*
> Find the following squares.
>
> | | | | | | | | |
> |---|---|---|---|---|---|---|---|
> | **(1)** | $4.671^2$ | **(2)** | $9.375^2$ | **(3)** | $14.87^2$ | **(4)** | $55.97^2$ |
> | **(5)** | $1105^2$ | **(6)** | $0.109^2$ | **(7)** | $0.023\ 41^2$ | **(8)** | $0.0081^2$ |

## 8.1  SQUARES

$7^2$    means $(7)(7)$

$\therefore 7^2 = 49$

$12^2$    means $(12)(12)$

$\therefore 12^2 = 144$

$4.61^2$ means $(4.61)(4.61)$

$\therefore 4.61^2 = 21.2521$

    or $= 21.25$ (rounded to 4 digits)

$356^2$ means $(356)(356)$

        $= 126\ 736$    or    $126\ 700$ (rounded to 4 digits)

**117**

## Practice Exercise No. 1

Find the following squares and round to 4 digits.

| | | | |
|---|---|---|---|
| **(1)** $21.9^2$ | **(2)** $89.2^2$ | **(3)** $16.8^2$ | **(4)** $1095^2$ |
| **(5)** $67.8^2$ | **(6)** $1.97^2$ | **(7)** $0.000\ 132^2$ | **(8)** $0.0373^2$ |
| **(9)** $0.139^2$ | **(10)** $0.953^2$ | **(11)** $137^2$ | **(12)** $162^2$ |
| **(13)** $89^2$ | **(14)** $951^2$ | **(15)** $104^2$ | **(16)** $99.44^2$ |
| **(17)** $1.04^2$ | **(18)** $775^2$ | **(19)** $0.0735^2$ | **(20)** $47.5^2$ |

---

*Self-Assessment Test No. 2*

Use square root tables to work out the following.

| | | | |
|---|---|---|---|
| **(1)** $\sqrt{47.81}$ | **(2)** $\sqrt{69.35}$ | **(3)** $\sqrt{10\ 190}$ | **(4)** $\sqrt{16\ 500}$ |
| **(5)** $\sqrt{0.951}$ | **(6)** $\sqrt{0.8751}$ | **(7)** $\sqrt{0.001\ 03}$ | **(8)** $\sqrt{0.0447}$ |

---

## 8.2 SQUARE ROOTS

$$\sqrt{49} = 7 \qquad \text{since } (7)(7) = 49$$
$$\sqrt{81} = 9 \qquad \text{since } (9)(9) = 81$$

For numbers that are not perfect squares, we need to use Table D1 in Appendix D. This table will give you the square root of 4-digit numbers between 1.000 and 9.999 on the first two pages and between 10.00 and 99.99 on the second two pages.

*Example (1)* To find $\sqrt{6.324}$.

    **(a)** Find row "6.3" in the left column of Table D1. Move over to column head "2." The junction of that row and column is 2.514, which is the square root of 6.32.

    **(b)** Column "4" of the proportional parts section of that row (the last three column groups, with single digit numbers) is 1, which is the amount to add to 2.514.

*Answer* $\sqrt{6.324} = 2.515$.

*Example (2)* To find $\sqrt{79.38}$.

    **(a)** Find row "79" and column "3" of Table D1 (since 79.34 is in the range 10.00 to 99.99). We find 8.905, which is the square root of 79.3.

    **(b)** Add the amount from the proportional parts section of that row under 8, which is 4, to 8.905.

*Answer* $\sqrt{79.38} = 8.909$.

If the number is larger than 100, or smaller than 1, we must "adjust" it before looking at the table. We do this by moving the decimal point, in pairs of two digits, left or right until we are in the range 1.000 to 99.99. That is, until we have either 1 or 2 digits to the left of the decimal point.

## Greater than 100

***Example (3)***  To find $\sqrt{6324.35}$.

(a) Move the decimal point (in pairs) to the left until 1 or 2 digits remain on the left of the decimal point.

$$6324.34$$

(b) Look up 63.24 (rounded to 4 digits) in Table D1.

$$\sqrt{63.24} = 7.953$$

(c) Move the decimal point back to the right *one digit for each pair* (one-half as many as before).

$$7.953$$

***Answer***    $\sqrt{6324.35} = 79.53$.

### Practice Exercise No. 2

Use Table D1 to work out the following.

| | | | |
|---|---|---|---|
| **(1)** $\sqrt{6.135}$ | **(2)** $\sqrt{9.661}$ | **(3)** $\sqrt{89.34}$ | **(4)** $\sqrt{16\,200}$ |
| **(5)** $\sqrt{6931}$ | **(6)** $\sqrt{877}$ | **(7)** $\sqrt{9510}$ | **(8)** $\sqrt{122\,500}$ |
| **(9)** $\sqrt{97.66}$ | **(10)** $\sqrt{101.96}$ | **(11)** $\sqrt{66.77}$ | **(12)** $\sqrt{444.6}$ |
| **(13)** $\sqrt{171.5}$ | **(14)** $\sqrt{19\,340}$ | **(15)** $\sqrt{1613}$ | **(16)** $\sqrt{97\,590}$ |

## Less than One

***Example (4)***  To find $\sqrt{0.9215}$.

(a) Move the decimal point (in pairs) to the right until 1 or 2 digits remain on the left of the decimal point.

$$0.9215$$

(b) Look up 92.15 in Table D1.

$$\sqrt{92.15} = 9.600$$

(c) Move the decimal point back *one place per pair*.

$$9.600$$

***Answer***    $\sqrt{0.9215} = 0.9600$.

***Example*** *(5)* To find $\sqrt{0.000\ 007\ 61}$.

    **(a)** Pair off (3 pairs).

$$0.000\ 007\ 61$$

    **(b)** Look up 7.61 in Table D1.

$$\sqrt{7.61} = 2.759$$

    **(c)** Move the decimal place back 3 places (add zeros as place holders).

$$0.0\,2.759$$

***Answer*** $\sqrt{0.000\ 007\ 61} = 0.002\ 759$.

## Practice Exercise No. 3

Use Table D1 to work out the following.

| | | | |
|---|---|---|---|
| **(1)** $\sqrt{0.19}$ | **(2)** $\sqrt{0.632}$ | **(3)** $\sqrt{0.197}$ | **(4)** $\sqrt{0.000\ 721\ 5}$ |
| **(5)** $\sqrt{0.837}$ | **(6)** $\sqrt{0.000\ 173\ 2}$ | **(7)** $\sqrt{0.85}$ | **(8)** $\sqrt{0.6}$ |
| **(9)** $\sqrt{0.132}$ | **(10)** $\sqrt{0.000\ 047\ 71}$ | **(11)** $\sqrt{0.0934}$ | **(12)** $\sqrt{0.047\ 61}$ |
| **(13)** $\sqrt{0.0875}$ | **(14)** $\sqrt{0.913}$ | **(15)** $\sqrt{0.0037}$ | **(16)** $\sqrt{0.5134}$ |

---

***Self-Assessment Test No. 3***

Use the theorem of Pythagoras to calculate the unknown sides in the following triangles.

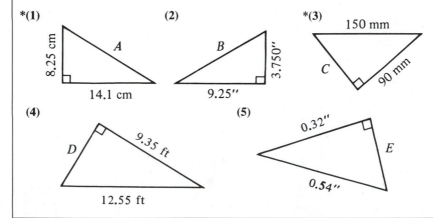

**\*(1)**    8.25 cm, A, 14.1 cm

**(2)**    B, 9.25″, 3.750″

**\*(3)**    150 mm, C, 90 mm

**(4)**    D, 9.35 ft, 12.55 ft

**(5)**    0.32″, E, 0.54″

# 8.3   THEOREM OF PYTHAGORAS

Pythagoras was a Greek mathematician. His theorem states that:

**In any right angled triangle, the square on the hypotenuse (long side) is equal to the sum of the squares on the two remaining sides.**

Thus, in the following triangle, $AB^2 = AC^2 + BC^2$.

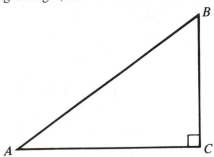

It is very useful to remember that triangles with sides of

3,  4,  5
5, 12, 13
7, 24, 25

and any multiples of these are right angled triangles.

---

**Calculator Hint**

Refer to Section C.4, Appendix C, on using a calculator to apply the Pythagorean theorem. Note: Your answers will vary slightly from the text in the examples that follow and in the exercises. Round your answer to three significant digits before comparing it to the text.

---

*\*Example (1)*   Find $a$.

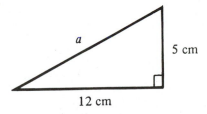

By Pythagoras,

$$a^2 = 5^2 + 12^2$$
$$= 25 + 144$$
$$= 169$$
$$a = \sqrt{169} = 13 \text{ cm}$$

*Example (2)*  Find $x$.

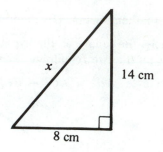

By Pythagoras,

$$x^2 = 8^2 + 14^2$$
$$= 64 + 196$$
$$= 260$$
$$x = \sqrt{260} = 16.1 \text{ cm}$$

*Example (3)*  Find $z$.

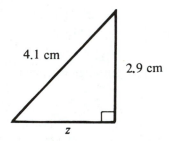

By Pythagoras, $2.9^2 + z^2 = 4.1^2$.
Subtracting $2.9^2$ from both sides,

$$z^2 = 4.1^2 - 2.9^2$$
$$= 16.81 - 8.41$$
$$= 8.40$$
$$z = \sqrt{8.40} = 2.90 \text{ cm}$$

*Example (4)*  Find $x$.

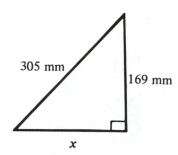

By Pythagoras, $x^2 + 169^2 = 305^2$
Subtracting $169^2$ from both sides,

$$x^2 = 305^2 - 169^2$$
$$= 93\ 000 - 28\ 600$$
$$= 64\ 400$$
$$x = \sqrt{64\ 400} = 254 \text{ mm}$$

## Practice Exercise No. 4

Find the missing lengths in the following triangles.

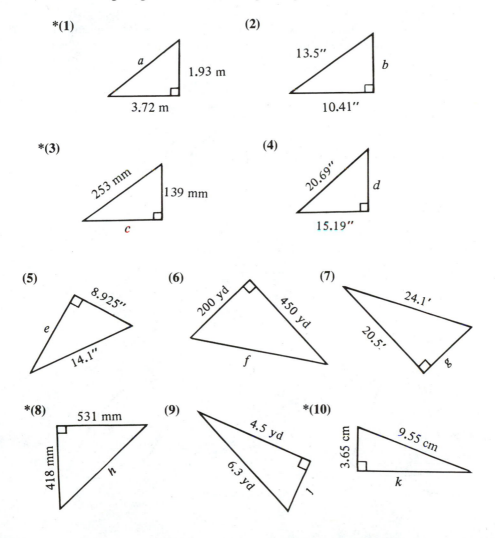

**\*(1)**

$a$   1.93 m   3.72 m

**(2)**

13.5″   $b$   10.41″

**\*(3)**

253 mm   139 mm   $c$

**(4)**

20.69″   $d$   15.19″

**(5)**

8.925″   $e$   14.1″

**(6)**

200 yd   450 yd   $f$

**(7)**

24.1′   20.5′   $g$

**\*(8)**

531 mm   418 mm   $h$

**(9)**

4.5 yd   6.3 yd   $j$

**\*(10)**

3.65 cm   9.55 cm   $k$

## 8.4  PRACTICAL APPLICATIONS
### EXERCISE

**(1)** Use the theorem of Pythagoras in Fig. 8–1 to calculate **(a)** *AE* and **(b)** *AC*.

**\*(2)** You are required to make a cone in Fig. 8–2, with base diameter (*AB*) 14 cm and perpendicular height (*CD*) 16 cm. The sides of the cone are to be cut from a flat plate, as shown in Fig. 8–3, and then shaped. Calculate the length of the slanting side (*l*).

**(3)** A square box is to be made to hold 6 ft³ of sand. What length of side is required if the box is to be 1 ft deep? (Volume = area of base × depth.)

**\*(4)** A pylon 49 m high is to be erected. It has two taut supporting ropes fixed to the top and these are secured in the ground 8 m and 10.5 m from the pylon as shown in Fig. 8–4. Calculate the lengths of support ropes *AB* and *CD*.

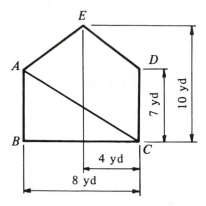

Fig. 8–1

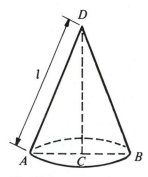

Fig. 8–2

Fig. 8–3

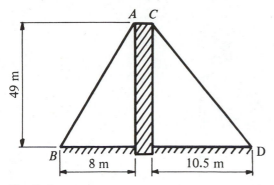

**Fig. 8–4**

**\*(5)** A lean-to garage is 3.5 m wide and the roof is pitched so that it is 3.7 m high on one side and 2.3 m high on the other (see Fig. 8–5). What is the rafter length (*l*) of the roof?

**(6)** A piece of *thin* wire is to be bent as shown in Fig. 8–6. Calculate the total length of wire required.

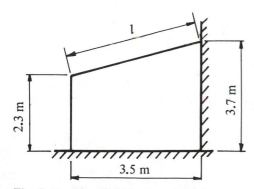

**Fig. 8–5**

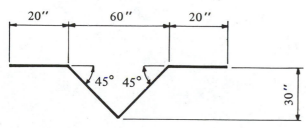

**Fig. 8–6**

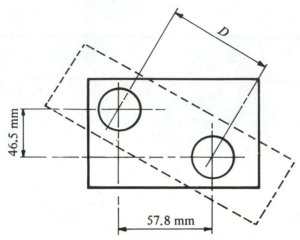

Fig. 8–7

(7) A telephone pole, 6.5 yd high, is situated 135 yd from a house. The pole is to be connected to the house by a wire running from the top of the pole to a point on the house 3.8 yd above ground. Assuming that both the pole and the house are on level ground, calculate the length of wire required.

*(8) A box measuring 71.5 cm by 44.9 cm (inside dimensions) is to be divided into two compartments by a *thin* diagonal partition. What is the length of this partition?

*(9) A plate is drilled as shown in Fig. 8–7. Another component (dotted line) is to be produced that incorporates two pins. These must be

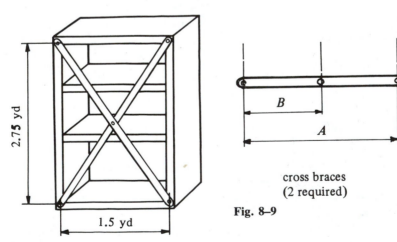

cross braces
(2 required)

Fig. 8–9

Fig. 8–8

located in the holes of the plate. From the dimensions given, calculate the distance $D$.

**(10)** A rack is to be erected as shown in Fig. 8–8. Unfortunately the unit, when assembled, is unstable and so two cross braces must be made (shown in position in Fig. 8–9). From the information given, calculate center distances $A$ and $B$.

**\*(11)** Use the theorem of Pythagoras to calculate **(a)** $x$ and **(b)** $y$ in Fig. 8–10.

**(12)** A casting is dimensioned as shown in Fig. 8–11; bores are indicated by shaded areas. From the information given, calculate center distances $A$, $B$, $C$, and $D$.

## 8.5  MISCELLANEOUS EXERCISES

**\*(1)** A square is to be machined on the end of a shaft as shown in Fig. 8–12. Calculate the maximum possible size of the square.

**(2)** Figure 8–13 shows the dimensions of the body of a dump truck. The body has to be strengthened by diagonal braces. Calculate the length of the diagonal $XY$.

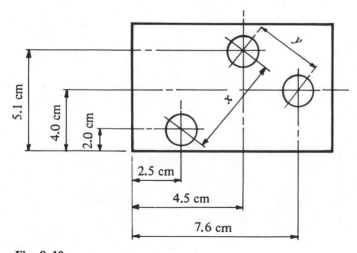

**Fig. 8–10**

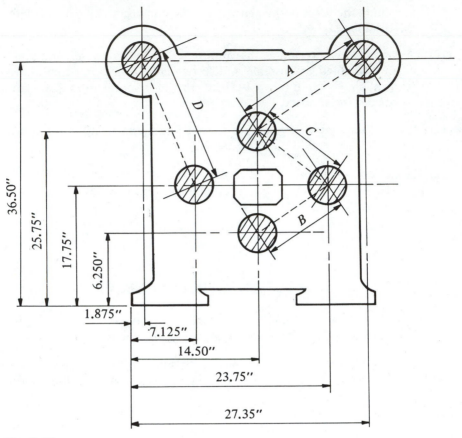

Fig. 8–11

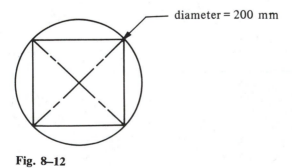

diameter = 200 mm

Fig. 8–12

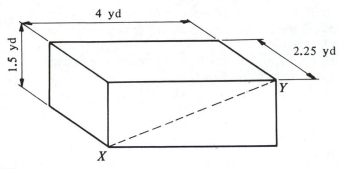

**Fig. 8–13**

*(3) Figure 8–14 shows a 10-mm flat on a 26-mm diameter bar. Calculate the depth of cut $h$.

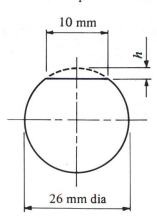

The sketch shows a 10 mm flat on a 26 mm diameter bar. Calculate the depth of cut $h$.

**Fig. 8–14**

(4) In Fig. 8–15, angle $ABC$ = angle $ACD$ = 90°. Calculate the lengths $AC$ and $CD$.

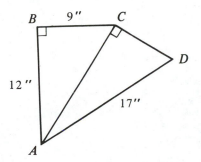

**Fig. 8–15**

*(5)   A wall fitting consists of lengths of wood joined as shown in Fig. 8–16. *ABCD* and *EFGH* are equal squares. *P* and *Q* are the midpoints of *BC* and *DC*, respectively.
   **(a)**   What is the total length of wood used to make the fitting?
   **(b)**   What is the distance from *A* to *G*?

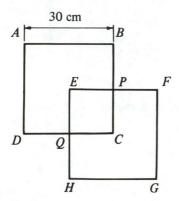

**Fig. 8–16**

*(6)   A component is to be machined as shown in Fig. 8–17. Calculate *x*.

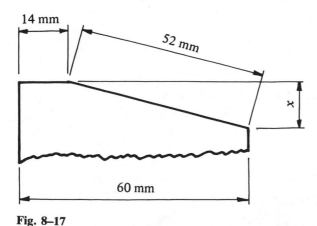

**Fig. 8–17**

(7)   A bracket is to be made as shown in Fig. 8–18. Calculate the total length of material required.

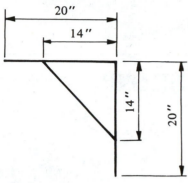

**Fig. 8–18**

# 9
# Areas of Plane Figures

*Self-Assessment Test No. 1*

Calculate the areas of the following shapes.

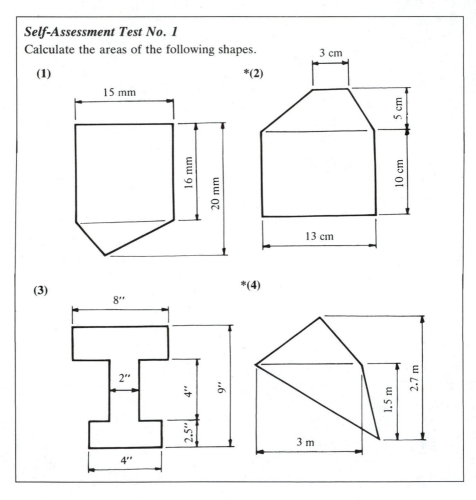

(1)

15 mm

16 mm

20 mm

*(2)

3 cm

5 cm

10 cm

13 cm

(3)

8″

2″

4″

2.5″

9″

4″

*(4)

1.5 m

2.7 m

3 m

132

> **Metrics**
>
> If you are unfamiliar with metric measurements of length (meters, centimeters, millimeters), you will find it helpful at this point to review Appendix B; pay particular attention to the discussions on length, area, and volume.

Area is a measure of a surface. It is measured in square units. For example, consider a carpet 3 m × 4 m.

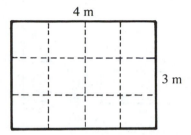

How many squares of side 1 m can be fitted into this shape?

$$Answer = 12$$

The area of the carpet is said to be 12 square meters or 12 m².

> **Note:**   *Units of area are m², in.², cm², mm², ft², yd², etc. You must always remember to include units in all calculations.*

## 9.1   COMMON PLANE FIGURES

Table 9–1 contains the formulas that may be used to find areas of common plane shapes.

**TABLE 9–1**

| Shape | Name & Description | Formula to Calculate Area |
|---|---|---|
| △ | **Triangle**<br><br>A three sided figure whose sides are straight lines. | area = $\frac{1}{2}$ base × perpendicular height<br><br>$A = \frac{1}{2} b h$ |

**TABLE 9–1 cont.**

| Shape | Name & Description | Formula to Calculate Area |
|---|---|---|
| | **Square**<br><br>All sides are equal, opposite sides are parallel. All angles are right angles. | area = base × perpendicular height<br><br>$A = b\,h$<br>But $b = h$<br><br>$\therefore\ \ A = b \times b = b^2$ |
| | **Rectangle**<br><br>Opposite sides are equal and parallel. All angles are right angles. | area = base × perpendicular height<br><br>$A = b\,h$ |
| | **Parallelogram**<br><br>Opposite sides are equal and parallel. | area = base × perpendicular height<br><br>$A = b\,h$ |
| | **Rhombus**<br><br>All sides are equal. Opposite sides are parallel. | area = base × perpendicular height<br><br>$A = b\,h$ |
| | **Trapezoid**<br><br>One pair of opposite sides are parallel. | area = $\frac{1}{2}$ (sum of parallel sides) × perpendicular height<br><br>$A = \frac{1}{2}\,(a + b)h$ |

**Note:** *Perpendicular means vertical or at right angles to the base.*

## Practice Exercise No. 1

Neatly copy and complete the following. ·

(1)  This figure is called a _____
Its area is given by the formula,
$A = $ _____ × _____
Its actual area,
$A = $ _____ × _____
    $= $ _____yd²

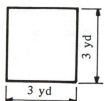

*(2)  This figure is called a _____
Its area is given by the formula,
$A = \dfrac{1}{2}$ _____ × _____
Its actual area,
$A = \dfrac{1}{2}$ _____ × _____
    $= $ _____cm²

(3)  This figure is called a _____
Its area is given by the formula,
$A = $ _____ × _____
Its actual area,
$A = $ _____ × _____
    $= $ _____in.²

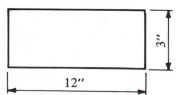

*(4)  This figure is called a _____
Its area is given by the formula,
$A = $ _____ × _____
Its actual area,
$A = $ _____ × _____
    $= $ _____mm²

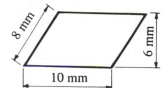

(5)  This figure is called a _____
Its area is given by the formula,
$A = $ _____ × _____
Its actual area,
$A = $ _____ × _____
    $= $ _____ft²

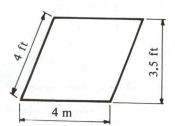

\*(6)  This figure is called a _____
Its area is given by the formula,

$$A = \frac{1}{2} (\underline{\hspace{1cm}}) \times \underline{\hspace{1cm}}$$

Its actual area,

$$A = \frac{1}{2} (\underline{\hspace{1cm}}) \times \underline{\hspace{1cm}}$$

$$= \underline{\hspace{1cm}} cm^2$$

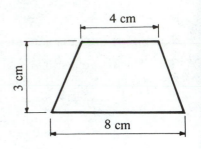

## Practice Exercise No. 2

Calculate the areas of the following shapes. (Sketch the shape and show all your work.)

**(1)**

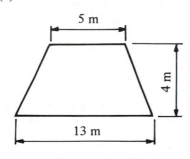

\*(2)

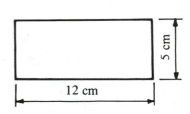

**(3)**

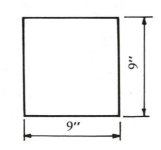

\*(4)

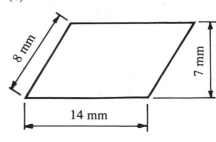

**(5)**

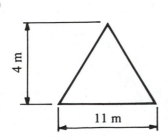

\*(6)

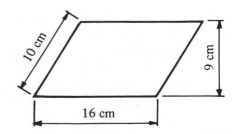

(7)    *(8)

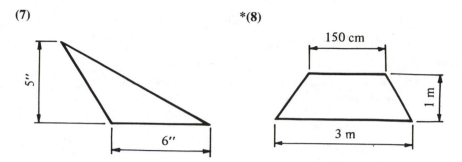

## 9.2   COMPOSITE FIGURES

Sometimes complicated shapes appear. These can be attempted as follows.

* *Example*   Find the area of the shape shown in Fig. 9–1.

The shape can easily be split into three rectangles $A$, $B$, and $C$.

$$\text{Area of } A = 30 \times 9 \text{ cm}^2 = \phantom{+}270 \text{ cm}^2$$
$$\text{Area of } B = 10 \times 9 \text{ cm}^2 = \phantom{+}\phantom{0}90 \text{ cm}^2$$
$$\text{Area of } C = 20 \times 6 \text{ cm}^2 = +120 \text{ cm}^2$$
$$\text{Total area} \phantom{= B } = \phantom{+}480 \text{ cm}^2$$

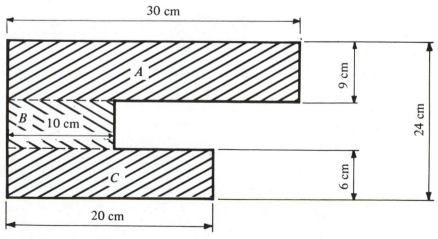

**Fig. 9–1**

## Practice Exercise No. 3

Copy and complete the following.

*(1)  Area of $A$ = ____ × ____ = _____ cm²
       Area of $B$ = ____ × ____ = _____ cm²
       Area of $C$ = ____ × ____ = _____ cm²
                    Total area = _____ cm²

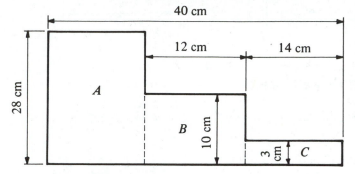

(2)  Area of shape = area of rectangle − area of pieces cut out
     Area of rectangle with no "cutouts" = ____ × ____ = _____ in.²
     Area of $B$ = ____ × ____ = _____ in.²
     Area of $C$ = ____ × ____ = _____ in.²

     Area of complete rectangle = ____ in.²
     Area of $(B + C)$          = ____ in.²
          Area of shape         = ____ in.²

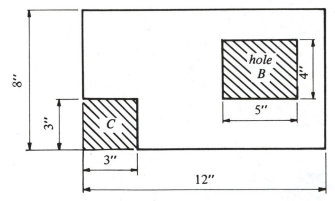

**\*(3)**   Area of triangle   $= \dfrac{1}{2}$ ___ × ___   = ___ cm²

Area of rectangle (without hole)

= ___ × ___   = ___ cm²

Area of trapezoid   $= \dfrac{1}{2}(\underline{\phantom{x}} + \underline{\phantom{x}}) \times$ ___ = ___ cm²

Total area of shape without hole   = ___ cm²

Area of hole   = ___ cm²

Area of shape   = ___ cm²

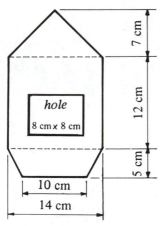

Calculate the areas of the following shapes.

**(4)**

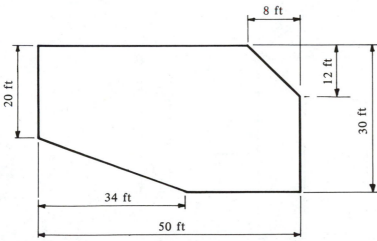

*(5)

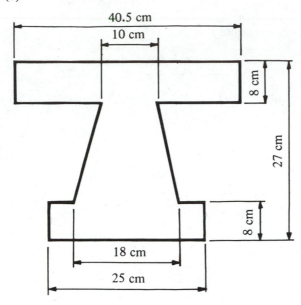

(6)

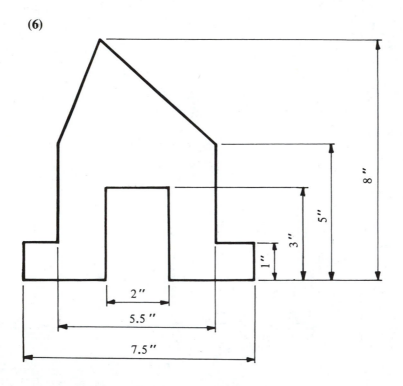

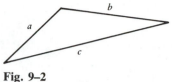

**Fig. 9–2**

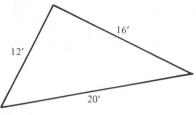

**Fig. 9–3**

## 9.3 HERO'S FORMULA FOR TRIANGLES

In the formula that we used to find the area of a triangle $\left( A = \dfrac{1}{2}bh \right)$, it was necessary to know the perpendicular height from one side (the base) to the opposite corner. That is not always practical, since it sometimes lies *inside* the triangle.

Another formula available uses only the lengths of the three sides. It is usually called **Hero's formula**. Consider Figure 9–2, with lengths of $a$, $b$, and $c$ for its sides. Using this figure, find

$$s = \frac{a + b + c}{2} \qquad \left( \frac{1}{2} \text{ of the perimeter} \right)$$

We can use the value of $s$ in Hero's formula to find the area.

---

**Hero's formula:**   Area $= \sqrt{s(s - a)(s - b)(s - c)}$

---

*Example (1)*   Find the area of Figure 9–3.

*Answer*   $s = \dfrac{12 + 16 + 20}{2} = \dfrac{48}{2} = 24$

$$\begin{aligned}
\text{Area} &= \sqrt{24(24 - 12)(24 - 16)(24 - 20)} \\
&= \sqrt{24(12)(8)(4)} \\
&= \sqrt{9216} \\
&= 96 \text{ ft}^2 \text{ (approx.)}
\end{aligned}$$

### Practice Exercise No. 4

Sketch and find the areas of the triangles with sides of the following lengths.

*(1)   8 cm, 8 cm, 8 cm

(2)   18″, 12″, 25″

*(3)   10 mm, 12 mm, 16 mm

(4)   100 ft, 120 ft, 180 ft

*(5)   2 m, 98 cm, 2 m

(6)   10 yd, 10 yd, 10 yd

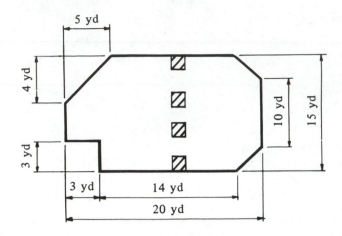

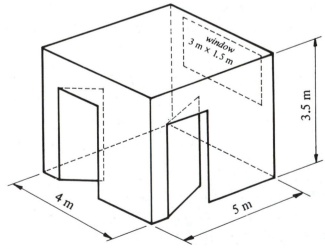

4 pillars each 1 yd X 1 yd

**Fig. 9–4**

## 9.4  PRACTICAL APPLICATIONS
## EXERCISE NO. 1

**(1)**  An apprentice builder has to calculate the floor area of a room as shown in Fig. 9–4. What would be his answer?

**\*(2)**  A plasterer has to plaster the walls of a kitchen. If the kitchen is as shown in Fig. 9–5, calculate the total surface area of the walls to be plastered. Doors are 1 m wide, 2 m high.

**Fig. 9–5**

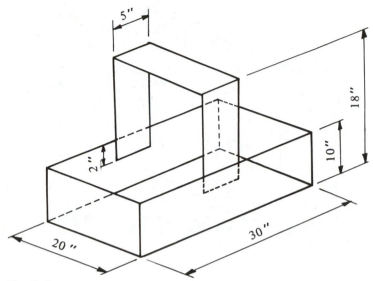

**Fig. 9–6**

(3)  A student designs a basket, as shown in Fig. 9–6, to be made out of thin sheet metal. Find the total surface area of material used. If the material costs 30¢ per 100 in.², find the total cost of material.

(4)  A room 6.5 yd by 4.5 yd is to be fitted with a new wooden floor using boarding that is 4.5″ wide. If 6.5-yd length boards are used, calculate the total length in yards of boarding required.

(5)  Calculate the number of bricks required to build the wall shown in Fig. 9–7. Allow $10\frac{2}{3}$ bricks per square foot of brickwork.

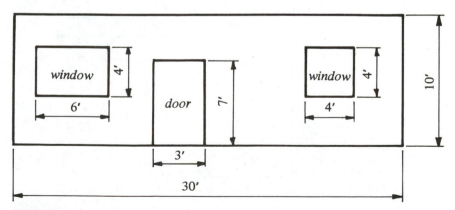

**Fig. 9–7**

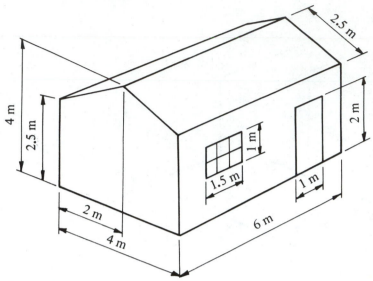

**Fig. 9–8**

**\*(6)** Find the cost of painting the outside of the building shown in Fig. 9–8 (including the roof). The window does not require painting. Take the cost of paint and labor as \$1.5/m² for sides and roof and \$1.35/m² for the door.

**(7)** A rectangular carton is designed as shown in Fig. 9–9. It is cut out and then folded along the broken lines and then glued. Calculate the actual surface area of material used.

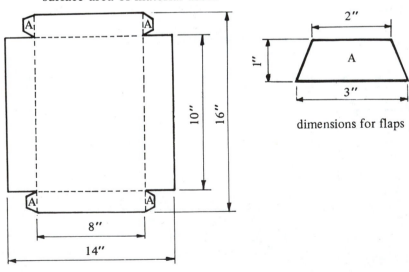

dimensions for flaps

**Fig. 9–9**

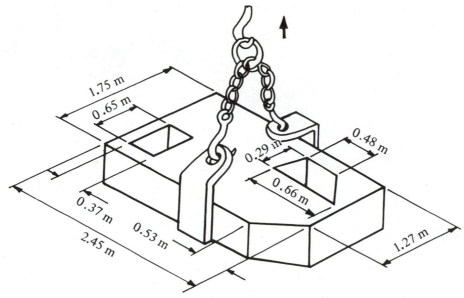

**Fig. 9–10**

(8) A gardener has to turf a lawn. The lawn is to completely cover the garden, which is 15 yd long and 14 yd wide, except for a path 1.3 yd wide all around it and an ornamental fish pond, 2 yd × 1.5 yd in the center.
 (a) Calculate the area to be turfed.
 (b) If the cost per square yard is $1.25, calculate the total cost of the turf used.

*(9) A component is made from material (of a certain thickness) whose mass is given as mass = 48.6 kg/m². The component has to be lifted by a crane (see Fig. 9–10).
 (a) Calculate the area of the top face in square meters.
 (b) Calculate the mass of the component. Mass = area of top face (m²) × 48.6 kg/m².

(10) The triangular lot in Fig. 9–11 is for sale. What is the area of the lot, in square yards?

(11) The triangle in Fig. 9–12 is the base of a triangular column supporting the portico of a sports arena. If the concrete flooring will support 284 pounds per square inch, what is the maximum weight that the column can exert on the flooring?

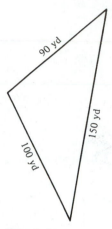

**Fig. 9–11**

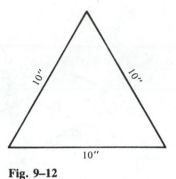

**Fig. 9–12**

---

*Self-Assessment Test No. 2*

*(1)   Convert 3200 mm² to cm².          *(2)   Convert 1.5 m² to cm².

(3)   Convert 12 ft² to in.².             (4)   Convert 156 in.² to ft².

(5)   Convert 9.63 ft² to yd².            (6)   Convert 16 310 in.² to ft².

*(7)   Convert 1 000 000 mm² to m².       *(8)   Convert 16 m² to cm².

---

## 9.5   CONVERTING SQUARE MEASUREMENTS

**\*To calculate the number of square centimeters in a square meter.**

The two squares shown here are identical (since 1 m = 100 cm).

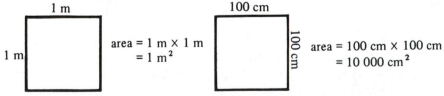

Because the two squares are the same size, 1 m² = 10 000 cm².

**To calculate the number of square inches in a square foot.**

The two squares shown here are identical (since 1 ft = 12″ or 12 in.).

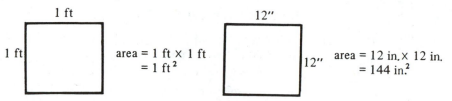

Because the squares are the same size, 1 ft² = 144 in.².

**To calculate the number of square feet in a square yard.**

The two squares shown here are identical (since 1 yd = 3 ft).

Because the squares are the same size, 1 yd² = 9 ft².

Using the above information it is possible to convert quantities of area from one type of unit to another.

***Example (1)***   Convert 4.5 m² to cm².

***Answer***   1 m² = 100 × 100 cm² = 10 000 cm²
$$\therefore 4.5 \text{ m}^2 = 4.5 \times 10\ 000 \text{ cm}^2$$
$$= 45\ 000 \text{ cm}^2$$

***Example (2)***   Convert 177 in.² to ft².

***Answer***   1 ft² = 12 × 12 in.² = 144 in.².

$$\therefore 1 \text{ in.}^2 = \frac{1}{144} \text{ ft}^2$$

$$177 \text{ in.}^2 = 177 \times \frac{1}{144} \text{ ft}^2 = \frac{177}{144} \text{ ft}^2$$

$$= 1.229 \text{ ft}^2$$

***Example (3)***   Convert 300 000 mm² to m².

***Answer***   1 m² = 1000 × 1000 mm² = 1 000 000 mm²

$$\therefore 1 \text{ mm}^2 = \frac{1}{1\ 000\ 000} \text{ m}^2$$

$$300\ 000 \text{ mm}^2 = \frac{300\ 000}{1\ 000\ 000} \text{ m}^2 = 0.3 \text{ m}^2$$

## Practice Exercise No. 5

**\*(1)** Convert the following from m² to cm².
    **(a)** 5 m²            **(b)** 7 m²            **(c)** 4.3 m²
    **(d)** 4.75 m²       **(e)** 4.63 m²

**\*(2)** Convert the following from cm² to m².
    **(a)** 40 000 cm²      **(b)** 110 000 cm²      **(c)** 780 000 cm²
    **(d)** 1100 cm²        **(e)** 120 cm²

**(3)** Convert the following from ft² to in.².
    **(a)** 2.76 ft²       **(b)** 3.67 ft²      **(c)** 8.81 ft²
    **(d)** 0.96 ft²       **(e)** 1.19 ft²

**(4)** Convert the following from in.² to ft².
    **(a)** 468 in.²       **(b)** 216 in.²      **(c)** 366 in.² (to 2 decimal places)

    **(d)** 998 in.² (to 2 decimal places)

## 9.6 PRACTICAL APPLICATIONS EXERCISE NO. 2

**(1)** A carpenter measures a board and finds that its area is 163 296 in.². The cost of material for the board is $9.50/yd². 
    **(a)** Calculate the area in square yards.
    **(b)** Calculate the total cost.

**\*(2)** An apprentice is told to make a pyramid from thin sheet metal. She finds that if a shape is cut out, as in Fig. 9–13 (shaded), and then folded along the broken lines, a pyramid will be produced, as in Fig. 9–14.

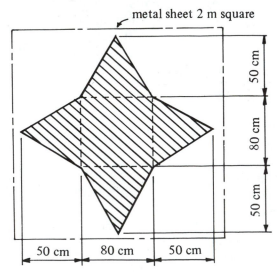

Fig. 9–13

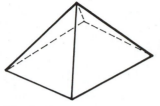

**Fig. 9–14**

    **(a)** Calculate the surface area of the pyramid, in square centimeters.
    **(b)** If the shape is to be cut from a sheet 2 m square, calculate how many square centimeters will be scrap.

*(3) When a metal bar is cut in half, its end is as shown in Fig. 9–15.
    **(a)** Calculate this area in square centimeters.
    **(b)** What is the area in square meters?

## 9.7   MISCELLANEOUS EXERCISES

*(1) Express 3.52 cm² in square millimeters.

(2) The shape of the gabled end wall of a building consists of a square and an isosceles triangle. The wall is 8 yd wide and 12 yd high from the base of the wall to the apex of the gable. Calculate the total area of the wall.

(3) Calculate the area, in square yards, of a rectangular floor measuring 2.5 yd by 4.5 yd.

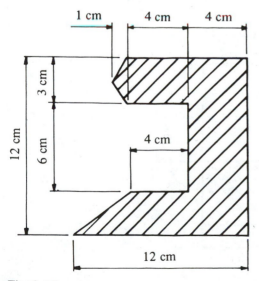

**Fig. 9–15**

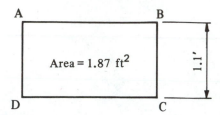

Fig. 9–16

  **\*(4)**  A rectangular windshield has a length of 1300 mm and a width of 350 mm.
      **(a)**  Calculate the area of the windshield.
      **(b)**  Determine the length of rubber strip required to seal the windshield.

  **\*(5)**  Calculate the area of a triangle of a vertical height of 50 cm and a base of 100 cm. Express your answer in square meters.

    **(6)**  Calculate the number of blocks required for a partition wall 20 ft long by 6 ft high. Allow 8 blocks per square yard.

  **\*(7)**  A ceiling 5.200 m long by 3.800 m wide is to be decorated with wood grain paper. The paper, allowing for wastage, covers 5.200 m² per roll. Calculate the number of rolls required for the job.

    **(8)**  **(a)**  If the area of the rectangle ABCD in Fig. 9–16 is 1.87 ft² and the width BC is 1.1 ft, what is the length AB?
      **(b)**  Express the area 1.87 ft² in square inches.

  **\*(9)**  The total surface area of a certain cube is 150 cm².
      **(a)**  How many faces has a cube?
      **(b)**  Find the length of an edge of the cube.

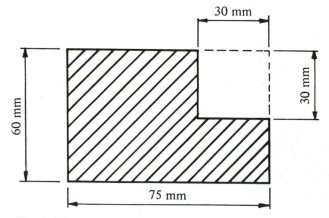

Fig. 9–17

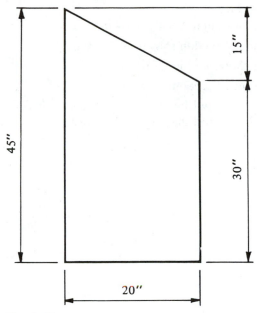

**Fig. 9–18**

(c)   Find how many such cubes can fit into a box 5 m × 2 m × $1\frac{1}{2}$ m.

**(10)**   A 26″ TV set measures 26″ across the diagonal of the screen, which is 20″ wide. Draw the screen to scale.

(a)   What is the height of the screen?

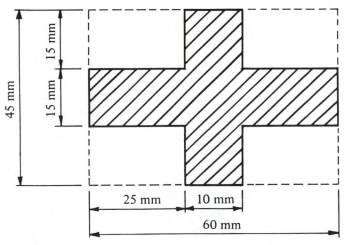

**Fig. 9–19**

    **(b)** What is the area of the screen?

    **(c)** What would the area of a similar 13″ TV screen?

*(11) Calculate the area of the shaded section shown in Fig. 9–17.

(12) Find the area of the section shown in Fig. 9–18.

*(13) Determine the area of the shaded section shown in Fig. 9–19.

(14) A rectangular carpet measures 16 yd by 20 yd and it has an inside border of 2 ft width all round it. Find the area of the border.

# 10

# The Circle and the Ellipse

*Self-Assessment Test No. 1*

Note that $\pi = 3.142$.

*(1)  Calculate the circumference of a circle of radius 4 cm.

*(2)  Calculate the circumference of a circle of diameter 3.5 m.

 (3)  Calculate the circumference of a circle of radius $9\frac{1}{2}''$.

 (4)  Calculate the circumference of a circle of diameter 45 ft.

Calculate the missing angles.

*(5)

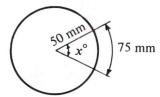

(6)

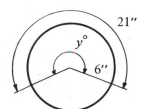

Calculate the missing lengths.

(7)

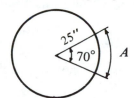

*(8)

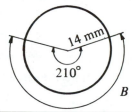

## 10.1   CIRCUMFERENCE

A **circle** is defined as "a curved line on which every point is equidistant from a point within, called the **center**."

**153**

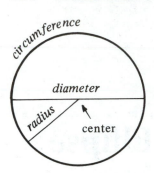

The **radius** of a circle is a straight line drawn from the center to the outer edge.

The **diameter** of a circle is a straight line drawn from any point on the outer edge, through the center, to the outer edge on the opposite side.

$$\text{Diameter} = 2 \times \text{Radius}$$
$$d = 2r \qquad \qquad \text{(Equation 1)}$$

The **circumference** of a circle is the complete distance around the circle.

One of the most interesting facts about the circle is the ratio of circumference to diameter. This ratio, $\dfrac{\text{circumference}}{\text{diameter}}$, is known as **Pi** and is written $\pi$. Its value is approximately $\dfrac{22}{7}$ or 3.142 correct to 3 decimal places. The value is the same for all circles.

$$\frac{\text{Circumference}}{\text{Diameter}} = \pi$$
$$\therefore \text{Circumference} = \pi \times \text{Diameter}$$
$$C = \pi d \qquad \qquad \text{(Equation 2)}$$

That is, to find the circumference of a circle, multiply its diameter by $\pi$ (3.142). From Equation 1, $d = 2r$. By substituting Equation 1 into Equation 2, we get

$$C = \pi 2r \qquad \text{or} \qquad C = 2\pi r$$

*Example (1)*   Calculate the circumference of the circle shown.

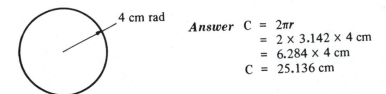

*Answer*   $C = 2\pi r$
$= 2 \times 3.142 \times 4 \text{ cm}$
$= 6.284 \times 4 \text{ cm}$
$C = 25.136 \text{ cm}$

*Example (2)* Calculate the circumference of the circle shown.

Answer $C = \pi d$
$= 3.142 \times 9.5''$
$C = 29.849''$

## Practice Exercise No. 1

Calculate the circumference of each of the following circles.

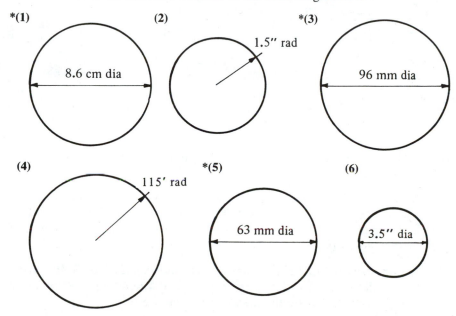

*(1)

8.6 cm dia

(2)

1.5″ rad

*(3)

96 mm dia

(4)

115′ rad

*(5)

63 mm dia

(6)

3.5″ dia

## 10.2 PRACTICAL APPLICATIONS
## EXERCISE NO. 1

*(1) A carpenter designs a table as shown in Fig. 10–1. He decides to use thin ornamental edging along the perimeter of the table top and also in three places on the central column. Calculate the length of edging required; give your answer in
(a) Centimeters.
(b) Meters.
(c) Millimeters.

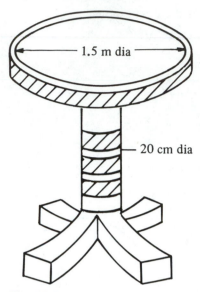

1.5 m dia

20 cm dia

**Fig. 10–1**

(2) If the radius of the earth is approximately 3964 miles, calculate the distance around the equator.

*(3) You are asked to make a dartboard. It consists of a fiber base with wire circles 34 cm, 32 cm, 20 cm, 18 cm, 3.5 cm, and 1.5 cm in diameter and straight pieces forming the scoring areas. Calculate the length of wire required (see Fig. 10–2).

(4) A woman wishes to make a lampshade as shown in Fig. 10–3. It is made from a flat piece of material and then folded around a metal frame, as in Fig. 10–4. Calculate the area of material used.

*(5) A well has a rope and bucket to draw water. How far will the bucket rise in 1 revolution of the handle? If the height of the drum center is 1 m and the distance from the drum center to the water is 31 m, estimate the number of turns required to raise the bucket from water level to the surface (see Fig. 10–5).

(6) An electrical instrument has a coil of 8000 turns of fine wire. If the average diameter of the coil is 6.5″, find the length of wire required in yards.

(7) If a wheel has a diameter of 70″, how far will it roll in 25 complete revolutions?

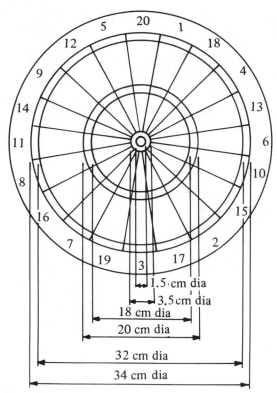

1.5·cm dia

3.5 cm dia

18 cm dia

20 cm dia

32 cm dia

34 cm dia

**Fig. 10–2**

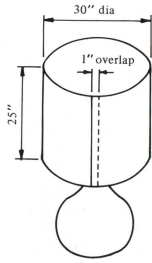

30″ dia

1″ overlap

25″

**Fig. 10–3**

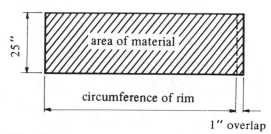

25″

area of material

circumference of rim

1″ overlap

**Fig. 10–4**

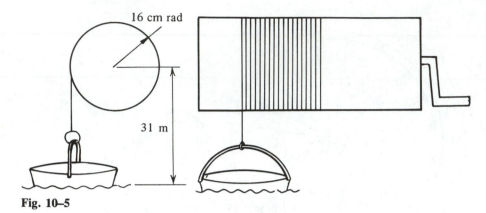

**Fig. 10–5**

## 10.3 CHORDS, ARCS, AND SECTORS

A **chord** is a straight line that joins two points on the circumference of a circle.

The chord $AB$ divides the circle into two **arcs**. $AXB$ is called the minor arc (the shorter one) and $AYB$ is called the major arc (the longer one).

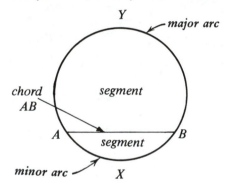

The chord $AB$ also divides the circle into two **segments**, which are the **areas** $AXB$ and $AYB$.

Another term commonly used is **sector**. This is the region bounded by an arc and two radii.

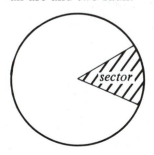

**To find the length of an arc of a circle that subtends a given angle at the center of the circle.**

*Example*    Find the length of arc *XY*. If the angle of the sector is increased, then *XY* will also increase. The length is dependent on the angle at the center. There are 360° in a circle.

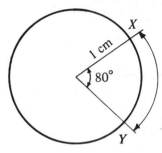

$$\text{Length of arc} = \text{Fraction of the circumference}$$

$$= \frac{\text{Sector angle}}{360} \times \text{Circumference}$$

In this case,

$$\text{Circumference} = 2\pi r$$

$$= 2 \times 3.142 \times 1 \text{ cm}$$

$$= 6.284 \text{ cm}$$

$$\therefore \text{Length of arc} = \frac{80}{360} \times 6.284$$

$$= 1.396 \text{ cm}$$

*Answer*    Arc *XY* = 1.396 cm (to 3 decimal places).

## Practice Exercise No. 2

*(1)   Calculate *AB*.         (2)   Calculate *CD*.         *(3)   Calculate *EF*.

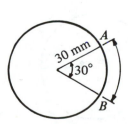

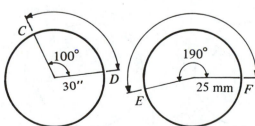

**(4)**   Calculate *GH*.        *(5)**   Calculate *IJ*.        **(6)**   Calculate *KL*.

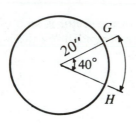 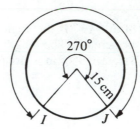 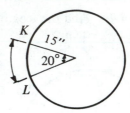

**To find the angle subtended at the center by an arc of given length.**

*Example*   Find *x*.

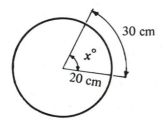

Angle *x* = Fraction of 360°

$$= \frac{\text{Arc length}}{\text{Circumference}} \times 360°$$

In this case,

$$\text{Circumference} = 2\pi r$$
$$= 2 \times 3.142 \times 20 \text{ cm}$$
$$= 125.68 \text{ cm}$$

$$\therefore x = \frac{30 \times 360}{125.68} = \frac{10\ 800}{125.68}$$

*Answer*   Angle *x* = 85.9° (to nearest tenth of a degree).

## Practice Exercise No. 3

Calculate the missing angles (to the nearest tenth of a degree).

*(1)                              *(2)                              (3)

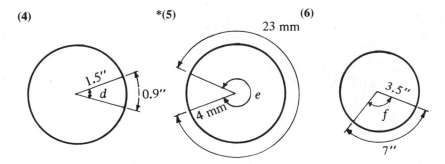

**(4)**       **\*(5)**       **(6)**

## 10.4  PRACTICAL APPLICATIONS
### EXERCISE NO. 2

**\*(1)**   A work surface is erected in a kitchen as shown in Fig. 10–6. A thin "edging" strip has to be fitted along the edges of the surface. Calculate the length required.

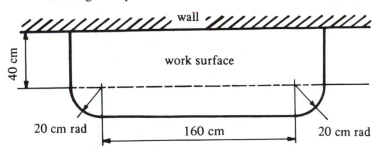

**Fig. 10–6**

**(2)**   Two identical pulleys are 1 yd apart. They are connected by a flat belt as shown in Fig. 10–7. Calculate the length of this belt.

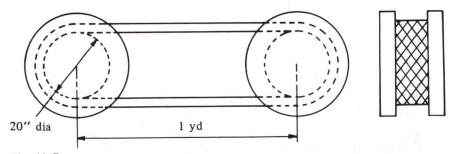

**Fig. 10–7**

**\*(3)**   The outside of a long semicircular drainage pipe, shown in Fig. 10–8, is to be painted. The can of paint to be used has written on its side "contents will cover 2 m²". Calculate the length of pipe that can be painted using this can of paint.

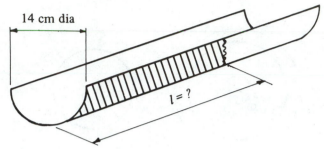

14 cm dia

l = ?

**Fig. 10–8**

*(4)   A retaining bracket for a securing bolt on a garden gate is shown in Fig. 10–9. It is made from a flat piece of material that is bent to shape. Assuming that the material is thin, calculate the length of material required to make this part.

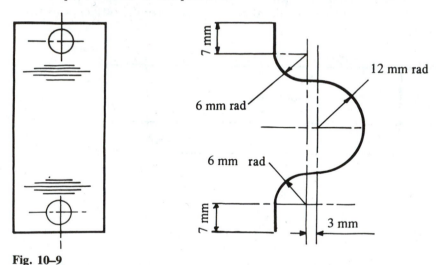

7 mm

6 mm rad

6 mm  rad

7 mm

12 mm rad

3 mm

**Fig. 10–9**

(5)   Two runners, $A$ and $B$, run once around the track shown in Fig. 10–10. Runner $A$ starts on the inside lane, and runner $B$ starts on the outside lane. If the two runners are to stay in their lanes throughout the race, calculate the stagger, $S$, needed in order that they will run the same distance.

(6)   In Fig. 10–11 a thin pipe is to be bent as shown from a straight length. Calculate the length of pipe required.

(7)   A thin pipe is to be bent as shown in Fig. 10–12. Calculate the length of pipe required before bending.

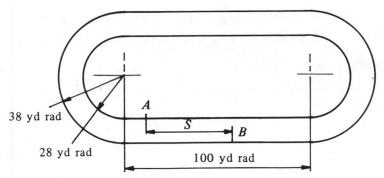

**Fig. 10–10**

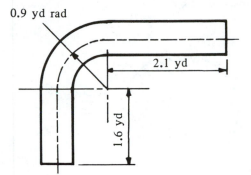

**Fig. 10–11**

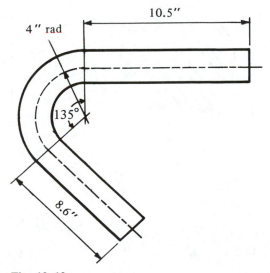

**Fig. 10–12**

(8)   A room consists of a rectangular part, plus a large semicircular bay on one end. The diameter of the bay is $2\frac{1}{2}$ yd. Calculate the length of baseboard required for the bay.

---

*Self-Assessment Test No. 2*

Note that $\pi = 3.142$.

*(1)   Calculate the area of a circle of radius 24 cm.

(2)   Calculate the area of a circle of diameter 150″.

*(3)   Calculate the area of a circle of radius 9.5 m.

(4)   Calculate the area of a circle of diameter 4.5 ft.

Find the areas of the following sectors.

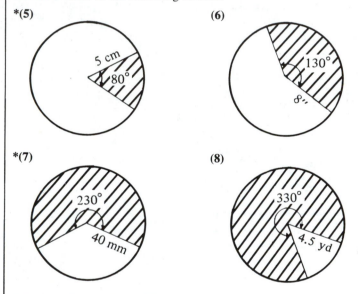

*(5)                                         (6)

*(7)                                         (8)

---

## 10.5   AREA OF A CIRCLE

The area ($A$) of a circle is given by the formula

$$A = \pi r^2 \qquad \text{(Equation 1)}$$

$$\text{Radius} = \frac{\text{Diameter}}{2} \quad \text{or} \quad r = \frac{d}{2} \qquad \text{(Equation 2)}$$

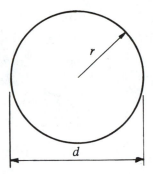

By substituting Equation 2 into Equation 1,

$$A = \pi r^2 = \pi \left(\frac{d}{2}\right)^2 = \pi \times \frac{d}{2} \times \frac{d}{2} = \pi \frac{d^2}{4}$$

$$\therefore \text{ Area of circle } = \pi r^2 \qquad \text{or} \qquad \pi \frac{d^2}{4}$$

***Example (1)***  Find the area of the following circle.

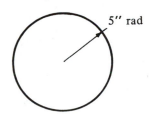

5″ rad

$$A = \pi r^2$$
$$= 3.142 \times 5 \times 5 \text{ in.}^2$$
$$= 3.142 \times 25 \text{ in.}^2$$
$$A = 78.55 \text{ in.}^2$$

***Example (2)***  Find the area of the following circle.

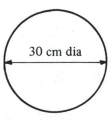

30 cm dia

$$A = \pi \frac{d^2}{4} = \frac{3.142 \times \overset{15}{\cancel{30}} \times \overset{15}{\cancel{30}}}{\underset{1}{\underset{2}{\cancel{4}}}} \text{ cm}^2$$

$$= 3.142 \times 15 \times 15 \text{ cm}^2$$
$$A = 706.95 \text{ cm}^2$$

## Practice Exercise No. 4

Find the areas of the following circles.

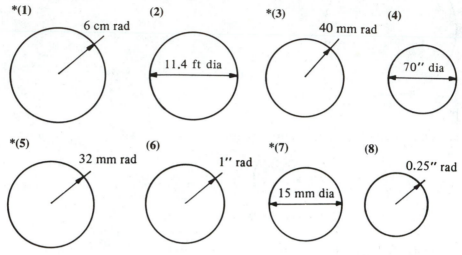

**\*(1)**    **(2)**    **\*(3)**    **(4)**

6 cm rad    11.4 ft dia    40 mm rad    70″ dia

**\*(5)**    **(6)**    **\*(7)**    **(8)**

32 mm rad    1″ rad    15 mm dia    0.25″ rad

### To find the area of a sector.

Find the area of sector $A$. The area of the sector $A$ is a fraction of the area of the circle.

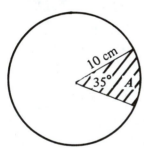

$$\text{Area of sector } A = \frac{35}{360} \times \text{Area of circle}$$

and,

$$\text{Area of circle} = \pi r^2$$
$$= \pi \times 10^2 = \pi \times 100 \text{ cm}^2$$
$$= 3.142 \times 100 \text{ cm}^2$$
$$\text{Area of circle} = 314.2 \text{ cm}^2$$
$$\therefore \text{ Area of sector} = \frac{35}{360} \times 314.2 \text{ cm}^2$$
$$= 30.55 \text{ cm}^2$$

### Practice Exercise No. 5

Find the areas of the following shaded sectors.

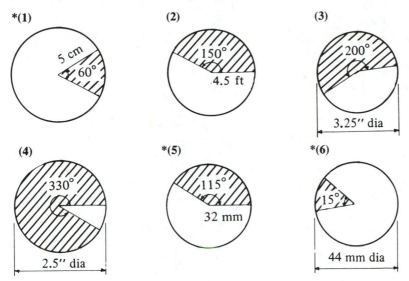

*(1)

(2)

(3)

5 cm 60°

150° 4.5 ft

200° 3.25″ dia

(4)

*(5)

*(6)

330° 2.5″ dia

115° 32 mm

15° 44 mm dia

## 10.6 PRACTICAL APPLICATIONS EXERCISE NO. 3

(1) A can is to be produced as illustrated in Fig. 10–13. It is made from three parts—a rectangular piece rolled into a tube plus two circular ends. Calculate the surface area of material used.

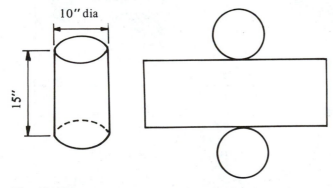

10″ dia

15″

**Fig. 10–13**

(2) Calculate the shaded area in Fig. 10–14.

(3) A builder is required to install a concrete floor as shown in Fig. 10–15. Calculate the floor area in square feet.

**Fig. 10–14**

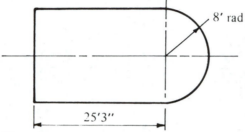

8' rad

25'3"

**Fig. 10–15**

**\*(4)** A gardener has a garden as shown in Fig. 10–16. Calculate the area of lawn.

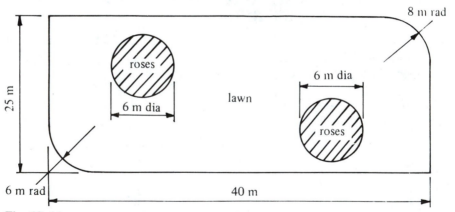

8 m rad

roses

6 m dia

lawn

6 m dia

roses

25 m

6 m rad

40 m

**Fig. 10–16**

**(5)** A sports arena is shown in Fig. 10–17. Calculate its area.

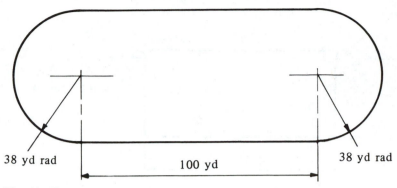

38 yd rad              100 yd              38 yd rad

**Fig. 10–17**

*(6)    A wall is shaped as shown in Fig. 10–18 and contains a window. Calculate the surface area of the wall.

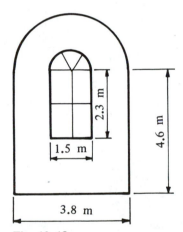

**Fig. 10–18**

*(7)    A circular baking dish is to be made as shown in Fig. 10–19. It is to be cut from flat plate and then bent to shape. Calculate the area of material used.

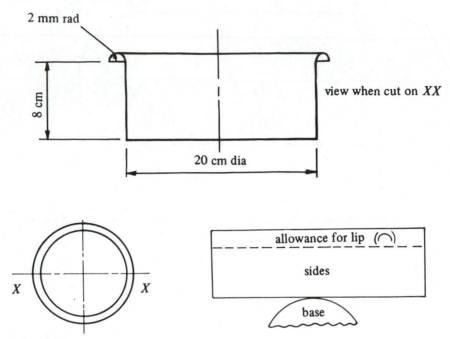

Fig. 10–19

## 10.7   THE ELLIPSE: AREA AND PERIMETER

The **ellipse**, shown in Fig. 10–20, is a figure similar to the circle, except that it has two diameters, called the **major axis** (or long diameter) and the **minor axis** (or short diameter).

The ellipse is occasionally seen in the construction trade in arches of bridges and in machinery in the form of elliptical gears and cams that vary the speed of motion.

The area and perimeter are calculated by using the following formulas.

$$a = \frac{1}{2} \text{ of the major axis} \qquad \text{and} \qquad b = \frac{1}{2} \text{ of the minor axis}$$

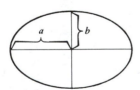

Fig. 10–20

## Area

$$\text{Area} = \pi ab$$

***Example (1)*** Find the area of an ellipse with axes of 10″ and 8″.

$$a = \frac{1}{2}(10) = 5'' \qquad b = \frac{1}{2}(8) = 4''$$

$$\begin{aligned}
\text{Area} &= \pi ab \\
&= 3.142(5)(4) \\
&= 62.84 \text{ in.}^2
\end{aligned}$$

## Perimeter

$$\text{Perimeter} = \pi\sqrt{2a^2 + 2b^2}$$

***\*Example (2)*** Find the perimeter of an ellipse with axes of 16 cm and 12 cm.

$$\begin{aligned}
\text{Perimeter} &= \pi\sqrt{2a^2 + 2b^2} \\
&= 3.142\sqrt{2(8)^2 + 2(6)^2} \\
&= 3.142\sqrt{2(64) + 2(36)} \\
&= 3.142\sqrt{128 + 72} \\
&= 3.142\sqrt{200} \\
&= 3.142\,(14.14) \\
&= 44.4 \text{ cm}
\end{aligned}$$

### Practice Exercise No. 6

Find the (a) area and (b) perimeter of each ellipse below.

|         | Minor axis | Major axis |
|---------|------------|------------|
| *(1)    | 10 cm      | 12 cm      |
| (2)     | 5″         | 6″         |
| *(3)    | 8 mm       | 15 mm      |
| (4)     | 28 ft      | 50 ft      |

## 10.8  MISCELLANEOUS EXERCISES

(1) Figure 10–21 shows a section of a circular steel shaft with a diameter of 80″ with a key-way of depth 20″ and a width of 25″. Calculate the cross-sectional area of the keyed shaft.

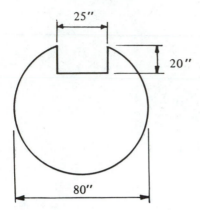

**Fig. 10–21**

**(2)** The cost of spraying a circular plate is 40¢ per 22″ of circumference. Find the cost of spraying a plate 70″ in diameter.

**\*(3)** Five 14-mm diameter holes are drilled in a plate. What is the total area of metal removed?

**(4)** A circular disc is 35″ in diameter. Calculate its circumference.

**(5)** Figure 10–22 shows the cross section of a circular tank of diameter $d$ and height $h$ with a flange of outside diameter $D$. Construct an equation for the surface area of the tank in terms of the symbols $d$, $D$, and $h$.

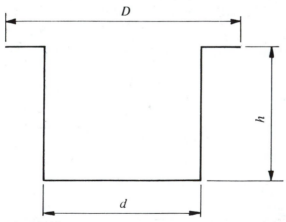

**Fig. 10–22**

**(6)** A wire-forming machine produces components as shown in Fig. 10–23. Calculate how many components can be produced from 50 yd of wire if 5% of the wire is wasted in cutting off the components. *(Finding the solution by measuring the drawing is not acceptable.)*

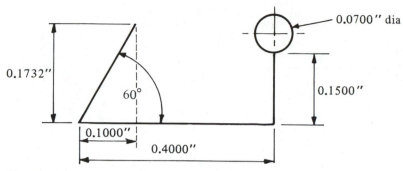

**Fig. 10–23**

*(7)   The cross-sectional area of an underground roadway is in the form of
a rectangle surmounted by a semicircle. The width of the roadway is
4 m and its maximum vertical height is 3.5 m. Determine the cross-
sectional area.

(8)   A cage attached to a rope is raised through a vertical mine shaft, 360
yd deep, when the rope is wound on a rotating, cylindrical drum of
diameter 3 yd. Find the number of revolutions made by the drum.

*(9)   The component shown in Fig. 10–24 is to be produced by press-working
methods. Calculate the perimeter of the component.

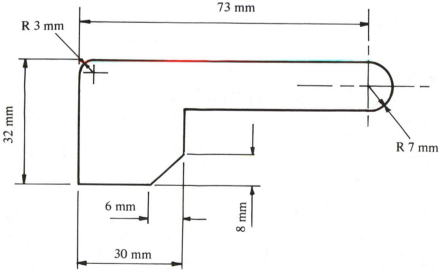

**Fig. 10–24**

*(10)   Calculate the perimeter of the small pressing shown in Fig. 10–25.

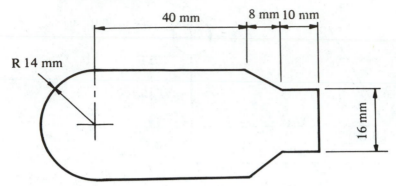

**Fig. 10–25**

*(11)   Figure 10–26 shows a template to be cut from sheet metal. Calculate the area of the template in square millimeters.

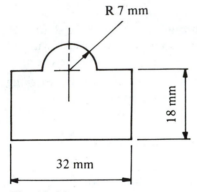

**Fig. 10–26**

(12)   A pipe has a diameter of 1″. What is its circumference?

(13)   A pipe support clip is shown in Fig. 10–27. What length of strip will be needed to make it?

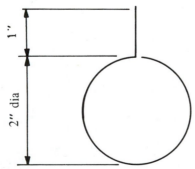

**Fig. 10–27**

**(14)** Calculate the flat length of material required to produce the component in Fig. 10–28.

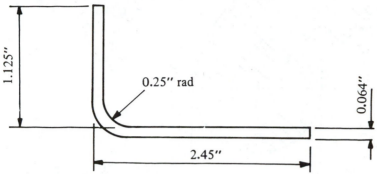

0.25" rad

1.125"

0.064"

2.45"

**Fig. 10–28**

**(15)** In Fig. 10–29,
    **(a)** Calculate the arc length $AB$.
    **(b)** Calculate the area of the sector $AOB$.

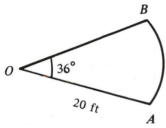

$B$

$O$ 36°

20 ft

$A$

**Fig. 10–29**

**(16)** $A$, $B$, and $C$ are the centers of the semicircles shown in Fig. 10–30. For the shaded region calculate **(a)** the perimeter and **(b)** the area. (Take $\pi \approx 3.14$.)

**(17)** **(a)** I have a circular lawn in my garden that has a radius of $3\frac{1}{2}$ yd. What is its area? $\left(\text{Take } \pi \approx \frac{22}{7}.\right)$

    **(b)** I can buy sods only in sets of 20. The area of each sod is $\frac{1}{3}$ yd². How many sods will I have to buy?

**\*(18)** Calculate the cost of insulating the 50-mm diameter pipe run shown in Fig. 10–31, using the following rates.
    **(a)** The 50-mm flexible foam pipe insulation at $3.00/m is sold in multiples of 1 m lengths only.

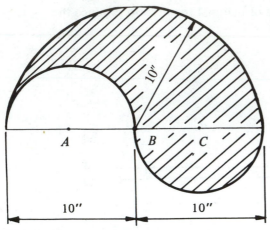

**Fig. 10–30**

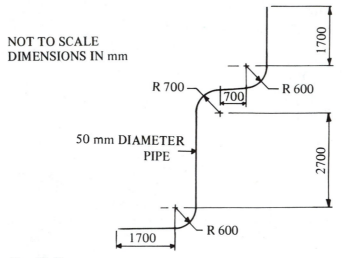

NOT TO SCALE
DIMENSIONS IN mm

**Fig. 10–31**

(b) Labor charges, including bends at $1.90/m, are to the nearest whole meter.

*(19) A duct of *circular* cross section, 700 mm in diameter, is to be replaced by a duct having the cross section shown in Fig. 10–32. If the replacement duct is to have the same cross-sectional area as the original (circular) duct, calculate its dimensions $R$ and $L$.

(20) An elliptical air duct with axes of 24″ and 12″ is to be replaced by a circular duct. If the ducts are to have the same cross-sectional area (in order to carry the same air pressure and volume), what should the diameter of the circular duct be?

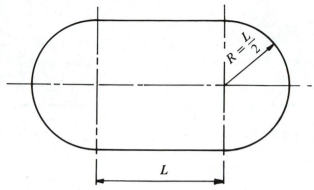

Fig. 10–32

# 11
# Solid Figures

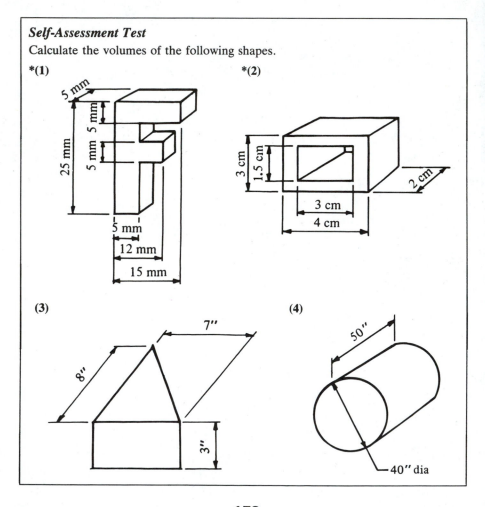

**Self-Assessment Test**

Calculate the volumes of the following shapes.

*(1)

*(2)

(3)

(4)

**178**

The measurement of space in three dimensions is called **volume**. Units of volume are cm³, mm³, m³, in.³, ft³, etc.

## 11.1   VOLUME

The space taken up by a cube of side 1 cm is 1 cm³, as shown in Fig. 11–1. Volume = 1 cm³.

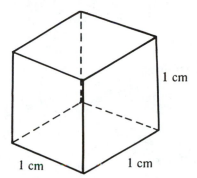

1 cm

1 cm          1 cm

**Fig. 11–1**

*Example*   What would be the volume of a block 8 cm × 4 cm × 1 cm (Fig. 11–2)? How many cubes of side 1 cm would fit into the shape?

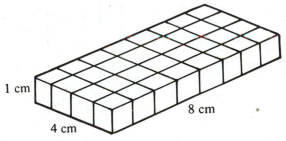

1 cm

4 cm

8 cm

**Fig. 11–2**

*Answer*   32

$$\therefore \text{Volume} = 32 \text{ cm}^3$$

The volume of a rectangular solid is given by $V = lwh$ (Fig. 11–3) where

$$l = \text{length}$$
$$w = \text{width}$$
$$h = \text{height}$$

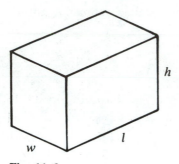

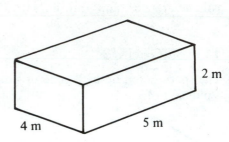

**Fig. 11–3**                    **Fig. 11–4**

***Example***   Find the volume of the block shown in Fig. 11–4.

$$\text{Volume} = lwh$$
$$= 5 \times 4 \times 2$$
$$= 40 \text{ m}^3$$

## Practice Exercise No. 1

Find the volume of the following solids.

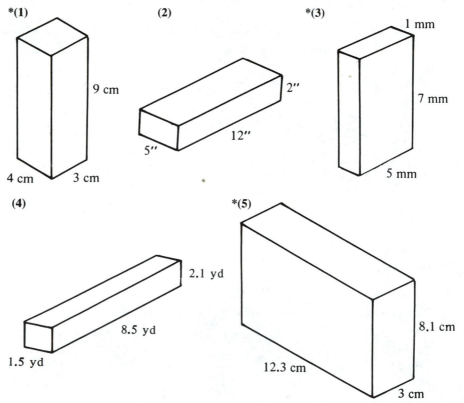

**\*(1)**                    **(2)**                    **\*(3)**

**(4)**                    **\*(5)**

## 11.2 CONVERTING CUBIC MEASUREMENTS

If the cube in Fig. 11–5 measures 1 m on all edges, then it also measures 100 cm on all edges (since 1 meter = 100 cm). In meters, its volume is 1 m × 1 m × 1 m = 1 m³. In centimeters, its volume is 100 cm × 100 cm × 100 cm = 1 000 000 cm³. Therefore,

$$1 \text{ m}^3 = 1\,000\,000 \text{ cm}^3$$

Similarly,

$$
\begin{aligned}
1 \text{ cm}^3 &= 1 \text{ cm} \times 1 \text{ cm} \times 1 \text{ cm} = 10 \text{ mm} \times 10 \text{ mm} \times 10 \text{ mm} \\
&= 1000 \text{ mm}^3 \\
1 \text{ ft}^3 &= 1 \text{ ft} \times 1 \text{ ft} \times 1 \text{ ft} = 12'' \times 12'' \times 12'' \\
&= 1728 \text{ in.}^3 \\
1 \text{ yd}^3 &= 1 \text{ yd} \times 1 \text{ yd} \times 1 \text{ yd} = 3 \text{ ft} \times 3 \text{ ft} \times 3 \text{ ft} \\
&= 27 \text{ ft}^3
\end{aligned}
$$

Therefore,

| |
|---|
| $1 \text{ m}^3 = 1\,000\,000 \text{ cm}^3$ |
| $1 \text{ cm}^3 = 1000 \text{ mm}^3$ |
| $1 \text{ ft}^3 = 1728 \text{ in.}^3$ |
| $1 \text{ yd}^3 = 27 \text{ ft}^3$ |

Using the above information, it is possible to convert quantities of volume from one type of unit to another.

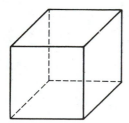

**Fig. 11–5**

*Example (1)*   Convert 4.5 m³ to cm³.

**Answer**   1 m³ = 1 000 000 cm³.
Therefore,

$$
\begin{aligned}
4.5 \text{ m}^3 &= 4.5 \times 1 \text{ m}^3 \\
&= 4.5 \times 1\,000\,000 \text{ cm}^3 \\
&= 4\,500\,000 \text{ cm}^3
\end{aligned}
$$

*Example (2)*    Convert 2700 mm³ to cm³.

*Answer*   1 cm³ = 1000 mm³.
Therefore,

$$\frac{1}{1000} \text{ cm}^3 = 1 \text{ mm}^3$$

$$2700 \text{ mm}^3 = 2700 \times 1 \text{ mm}^3$$

$$= 2700 \times \frac{1}{1000} \text{ cm}^3$$

$$= 2.700 \text{ cm}^3$$

*Example (3)*    Convert 278 ft³ to in.³.

*Answer*   1 ft³ = 1728 in.³.
Therefore,

$$278 \text{ ft}^3 = 278 \times 1 \text{ ft}^3$$

$$= 278 \times 1728 \text{ in.}^3$$

$$= 480\ 384 \text{ in.}^3$$

*Example (4)*    Convert 120 ft³ to yd³.

*Answer*   1 yd³ = 27 ft³.
Therefore,

$$\frac{1}{27} \text{ yd}^3 = 1 \text{ ft}^3$$

$$120 \text{ ft}^3 = 120 \times 1 \text{ ft}^3$$

$$= 120 \times \frac{1}{27} \text{ yd}^3$$

$$= \frac{120}{27} \text{ yd}^3$$

$$= 4.44 \text{ yd}^3$$

## Practice Exercise No. 2

Convert the following units into those specified.

(1)   351 yd³ = _____ ft³

*(2)   25 m³ = _____ cm³

(3)   375 ft³ = _____ in.³

*(4)   275 385 cm³ = _____ m³

(5)   38 ft³ = _____ yd³

*(6)   23 127.85 cm³ = _____ m³

(7)   12 758 in.³ = _____ ft³

*(8)   3758.3 cm³ = _____ mm³

## 11.3   GENERAL FORMULAS
##         FOR VOLUMES AND
##         SURFACE AREAS OF PRISMS

A **prism** is a solid figure with parallel edges and a uniform cross section. In other words, if you have a solid shape and cut it parallel to the base, in various places, then the area of the cut face will be the same no matter where you cut the shape.

Note that, in this definition, the base may not be the face on which the solid is standing. Thus, in the case of the triangular prism shown in Fig. 11–7, either of the faces P or Q can be taken as the base.

Consider the prisms shown in Figs. 11–6 and 11–7. If Fig. 11–6 is cut at *A* and *B*, the areas of the cut faces will be different. But if it is cut at *C* and *D* (as in Fig. 11–7) then the areas of the cut faces will be the same. This constant area is called the **cross-sectional area**. The area of the three rectangles making the sides of the prism is called the **lateral surface area** (**LSA**) or wraparound area.

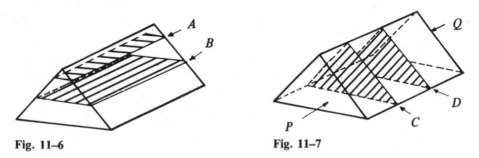

Fig. 11–6                                Fig. 11–7

If the edges of the prism are at right angles to the base, the prism is called a **right prism**. When the base of a right prism is a circle, the figure is called a **cylinder**, as in Fig. 11–8. That is, in a cylinder the sides are parallel and at right angles to the base. Horizontal sections are all equal circles.

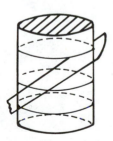

Fig. 11–8

Thus, we have the following two formulas.

Volume of a prism = Cross-sectional area × Length

where length = length of parallel sides

Lateral surface area = Perimeter of cross-sectional area × Length

### Practice Exercise No. 3

Look at the following shapes and state which are prisms (i.e., which have uniform cross section).

**(1)**                **(2)**                **(3)**

**(4)**                        **(5)**

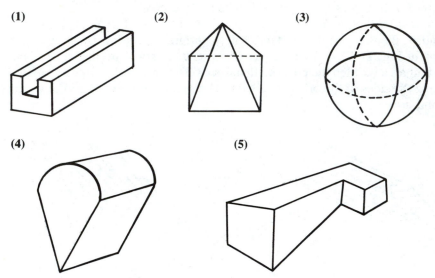

## Calculation of Volumes and Surface Areas of Prisms

*Example (1)*  **(a)**  Find the volume of the shape in Fig. 11–9.

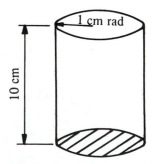

**Fig. 11–9**

$$\text{Cross-sectional area} = \pi r^2$$
$$= 3.142 \times 1^2$$
$$= 3.142 \text{ cm}^2$$

$$\text{Volume} = \text{Cross-sectional area} \times \text{Length}$$
$$= 3.142 \times 10$$
$$= 31.42 \text{ cm}^3$$

**(b)** Find the surface area of Fig. 11–9.

*Answer*   Perimeter (Circumference) $= 2\pi r = \pi d$.
$$= 3.142 \times 2 \text{ cm}$$
$$= 6.284 \text{ cm}$$

$$\text{Lateral surface area (LSA)} = \text{Perimeter} \times \text{Length}$$
$$= 6.284 \text{ cm} \times 10 \text{ cm}$$
$$= 62.84 \text{ cm}^2$$

$$\text{Total surface area (TSA)} = \text{LSA} + 2 \times \text{Cross-sectional area}$$
$$= 62.84 \text{ cm}^2 + 2 \times 3.142 \text{ cm}^2$$
$$= 62.84 \text{ cm}^2 + 6.284 \text{ cm}^2$$
$$= 69.124 \text{ cm}^2$$

*Example (2)*   Find the volume of the following shape.

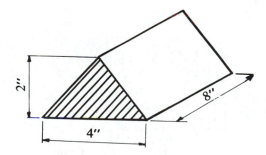

$$\text{Cross-sectional area} = \frac{1}{2} \times 4 \times 2 = 4 \text{ in.}^2$$

$$\text{Volume} = \text{Cross-sectional area} \times \text{Length}$$
$$= 4 \times 8$$
$$= 32 \text{ in.}^3$$

*Example (3)*  Find the volume of the following shape.

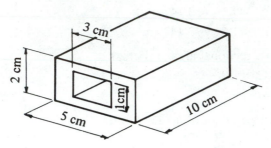

Cross-sectional area $= (2 \times 5) - (1 \times 3)$

$$= 10 - 3$$
$$= 7 \text{ cm}^2$$

Volume $=$ Cross-sectional area $\times$ Length

$$= 7 \times 10$$
$$= 70 \text{ cm}^3$$

## Practice Exercise No. 4

Find the volumes and the LSA of the following where indicated.

*(1)  Volume and LSA.

(2)  Volume and LSA.

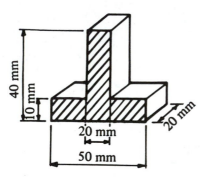

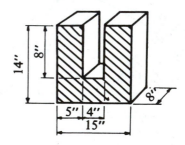

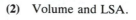

*(3)  Volume and TSA.

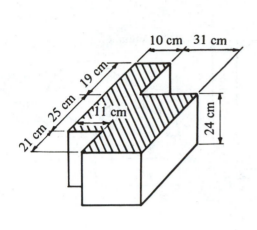

(4)  Volume.

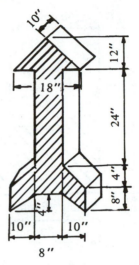

*(5)  Volume and TSA.

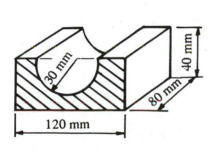

(6)  Volume.

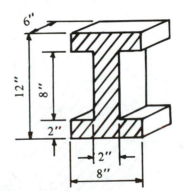

*(7)  Volume.

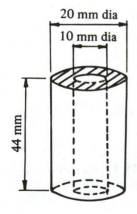

(8)  Volume and LSA.

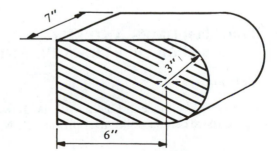

*(9) Volume.

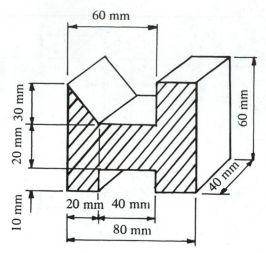

(10) Volume.

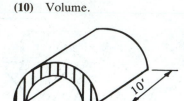

(11) Volume (cross section is a regular hexagon). The area of a regular hexagon is $2.598s^2$, where $s$ is the length of a side.

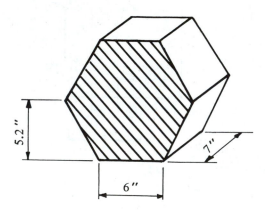

## 11.4 VOLUMES AND SURFACE AREAS OF PYRAMIDS AND CONES

### Volume

Pyramids, like those in Figs. 11–10, 11–11, and 11–12, are formed by a plane figure (base) and a point above it (apex). Notice that the sides are triangles.

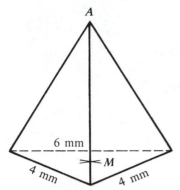

6 mm

4 mm    4 mm

×M

**Fig. 11–10**

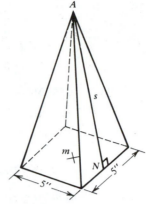

A

s

m×

N

5"

5"

**Fig. 11–11**

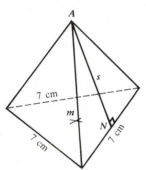

A

7 cm

s

m×

N

7 cm

7 cm

**Fig. 11–12**

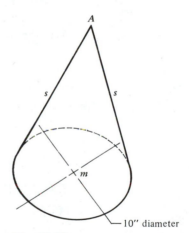

A

s         s

×m

10" diameter

**Fig. 11–13**

Cones, like Fig. 11–13, are formed by a circle (base) and a point above it (apex).

To find the volumes of these four solids, we need the area of the base and the altitude (perpendicular distance) from the base to the apex, $A$ to $M$, in each of these figures. Notice that $M$ does not necessarily have to be the center of the base, just directly under the apex.

$$\text{Volume} = \frac{(\text{Area of the base}) \times (\text{Altitude})}{3}$$

**\*Example (1)**   Find the volume of Fig. 11–10 if $\overline{AM}$ = 10 mm.

**Answer**   First we find the area of the triangular base using Hero's formula.

$$s = \frac{(6 + 4 + 4)}{2} = \frac{14}{2} = 7$$

$$\text{Area} = \sqrt{7(7 - 6)(7 - 4)(7 - 4)}$$

$$= \sqrt{7(1)(3)(3)}$$

$$= \sqrt{63}$$

$$= 7.94 \text{ mm}^2$$

$$\text{Volume} = \frac{7.94 \times 10}{3}$$

$$= 26.47 \text{ mm}^3$$

*Example (2)*   Find the volume of Fig. 11–11 if $\overline{AM}$ = 10".

*Answer*   Area of square base = 5" × 5"

$$= 25 \text{ in.}^2$$

$$\text{Volume} = \frac{25 \times 10}{3}$$

$$= 83.33 \text{ in.}^3$$

*\*Example (3)*   Find the volume of Fig. 11–12 if $\overline{AM}$ = 12 cm.

*Answer*   Find the area of triangular base.

$$s = \frac{(7 + 7 + 7)}{2} = \frac{21}{2} = 10.5 \text{ cm}$$

$$\text{Area} = \sqrt{10.5(10.5 - 7)(10.5 - 7)(10.5 - 7)}$$

$$= \sqrt{10.5(3.5)(3.5)(3.5)}$$

$$= \sqrt{450.12}$$

$$= 21.2 \text{ cm}^2$$

$$\text{Volume} = \frac{21.2 \times 12}{3}$$

$$= 84.8 \text{ cm}^3$$

*Example (4)*   Find the volume of Fig. 11–13 if $\overline{AM}$ = 12".

*Answer*   Find the area of circular base.

$$\text{Area} = \pi r^2$$

$$= 3.142 \times (5)^2$$

$$= 78.55 \text{ in.}^2$$

$$\text{Volume} = \frac{78.55 \times 12}{3}$$

$$= 314.2 \text{ in.}^3$$

## Surface Areas

The lateral surface area of a **pyramid** is the sum of the areas of the triangular faces. This is difficult to find if the faces are all different, as in Fig. 11–10. However, if the faces are identical, we may use the following formula.

$$\text{Lateral surface area} = \frac{1}{2} (\text{perimeter of base})(\text{slant height})$$

where

$$\text{Slant height} = \text{perpendicular distance from the \textbf{apex} to a \textbf{side} of the base}$$

This formula also applies to **cones**.

> **Note:** *In order to use the above formula, all faces must be identical. That requires that:*
>
> *(1) The apex be above the center of the base.*
>
> *(2) The base be a regular figure (all sides are equal).*
>
> *Such a pyramid or cone is called a **right pyramid** or **right cone**.*

*Example (5)* Find the total surface area of Fig. 11–11 if the altitude $\overline{AM}$ is 10″.

**Answer**

> The slant height is the hypotenuse of triangle $AMN$; thus,
>
> $$s^2 = (\overline{AM})^2 + \left(\frac{1}{2} \text{ of distance from side to side}\right)^2$$
> $$= (10″)^2 + \left(\frac{1}{2} \text{ of } 5″\right)^2$$
> $$= 100 + 6.25$$
> $$\therefore s = \sqrt{106.25} = 10.31″$$

This process of finding the **slant height** works on most regular pyramids with an even number of sides (2, 4, 6, etc.) on the base.

$$\text{Perimeter} = 5″ + 5″ + 5″ + 5″ = 20″$$

$$\text{LSA} = \frac{1}{2} \times 20″ \times 10.31″$$

$$= 103.1 \text{ in.}^2$$

$$\text{Total surface area} = \text{LSA} + \text{Base area}$$
$$= 103.1 \text{ in.}^2 + 25 \text{ in.}^2$$
$$= 128.1 \text{ in.}^2$$

*Example (6)*  Find the lateral surface area of Fig. 11–12, if the slant height $s$ is 12.17 cm (it is not as easy to calculate this using the height and sides, as in Fig. 11–11, since Fig. 11–12 has an odd number of sides on the base).

*Answer*

$$\text{Perimeter} = 7 \text{ cm} + 7 \text{ cm} + 7 \text{ cm} = 21 \text{ cm}$$

$$\text{LSA} = \frac{1}{2} \times 21 \text{ cm} \times 12.17 \text{ cm}$$

$$= 127.8 \text{ cm}^2$$

**Example (7)**  Find the lateral surface area of Fig. 11–13, if $\overline{AM} = 12''$.

*Answer*

$$\text{Slant height: } s^2 = 12^2 + \left(\frac{1}{2} \text{ of } 10''\right)^2$$
$$= 12^2 + 5^2$$
$$= 144 + 25$$
$$s^2 = 169$$
$$s = \sqrt{169}$$
$$s = 13''$$

$$\text{Perimeter (circumference)} = \pi d$$
$$= 3.142 \times 10''$$
$$= 31.42''$$

$$\text{LSA} = \frac{1}{2} \times 31.42'' \times 13''$$

$$= 204.23 \text{ in.}^2$$

## Practice Exercise No. 5

Find the volume and lateral surface area of the following solids.

**(1)**  A right pyramid whose base is a square with sides of 6″ and an altitude of 8″.

**(2)**  A right cone of diameter 8″ and altitude of 10″.

**(3)**  A right pyramid whose base is a square 10″ by 10″, and whose altitude is 5″.

*(4)  A right cone whose base is a 15-mm radius circle and whose altitude is 25 mm.

Find the volume of the following figures.

(5)  A pyramid whose base is a 10″ by 8″ rectangle and whose altitude is 5″.

*(6)  A "cone" whose base is an ellipse with axes of 6 cm and 4 cm and an altitude of 5 cm.

## 11.5  FRUSTUMS OF PYRAMIDS AND CONES

A **frustum** of a pyramid or cone is formed by cutting the top off with a plane parallel to the base. This results in a plane figure on the top with the same shape as the base, as in Figs. 11–14 to 11–18.

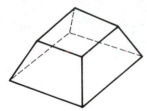

Fig. 11–14

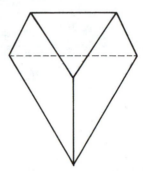

Fig. 11–15

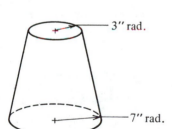

3″ rad.

7″ rad.

Fig. 11–16

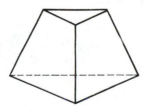

Fig. 11–17

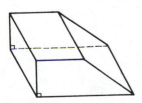

Fig. 11–18

## Volume

If
$$
\begin{aligned}
T &= \text{area of the top} \\
B &= \text{area of the base} \\
a &= \text{altitude} \quad \text{(perpendicular distance} \\
&\qquad\qquad\qquad \text{from top to base)}
\end{aligned}
$$

$$
\text{Volume} = \frac{a(B + T + \sqrt{BT})}{3}
$$

*Example (1)*  Find the volume of Fig. 11–14 if the base is a square, 5 ft by 5 ft, the top is a square, 3 ft by 3 ft, and the height is 3 ft.

*Answer*

$$
B = 5 \times 5 = 25 \text{ ft}^2
$$
$$
T = 3 \times 3 = 9 \text{ ft}^2
$$
$$
a = 3 \text{ ft}
$$

$$
\text{Volume} = \frac{3(25 + 9 + \sqrt{25 \times 9})}{3}
$$

$$
\text{Volume} = \frac{3(25 + 9 + \sqrt{225})}{3}
$$

$$
= \frac{3(25 + 9 + 15)}{3}
$$

$$
= 49 \text{ ft}^3
$$

*Example (2)*  Find the volume of Fig. 11–16 if the height of the frustum is 20″.

*Answer*

$$
\begin{aligned}
B &= \pi r^2 \\
&= 3.142(7)^2 \\
&= 154.0 \text{ in.}^2 \\
T &= \pi r^2 \\
&= 3.142(3)^2 \\
&= 28.3 \text{ in.}^2
\end{aligned}
$$

$$
\text{Volume} = \frac{20(154.0 + 28.3 + \sqrt{154.0 \times 28.3})}{3}
$$

$$
= \frac{20(154.0 + 28.3 + 66.0)}{3}
$$

$$
= \frac{20(248.3)}{3}
$$

$$
= 1655.3 \text{ in.}^3
$$

*Example (3)*  Find the volume of Fig. 11–18 if the top is a 12 cm by 18 cm rectangle, the base is a 24 cm by 36 cm rectangle, and the altitude is 9 cm.

*Answer*

$$T = 12 \times 18 = 216 \text{ cm}^2$$
$$B = 24 \times 36 = 864 \text{ cm}^2$$
$$a = 9 \text{ cm}$$

$$\text{Volume} = \frac{9(864 + 216 + \sqrt{216 \times 864})}{3}$$

$$= \frac{9(1512)}{3}$$

$$= 4536 \text{ cm}^3$$

## Surface Area

Lateral surface area is the sum of the faces between the top and bottom, which within the frustum of a pyramid are trapezoids.

For **right pyramids** and **right cones** the **lateral surface** of the **frustum** depends on the average perimeter of the top and base.

$$\text{Average perimeter} = \frac{(\text{Perimeter of top} + \text{Perimeter of base})}{2}$$

$$\text{Lateral surface area} = (\text{Average perimeter})(\text{Slant height})$$

Note that Figs. 11–17 and 11–18 cannot be worked with this formula since they are not right pyramids.

*Example (4)*  Find the total surface area of Fig. 11–14 if the top is a 3 ft square, the base is a 5 ft square, and the altitude is 3 ft.

*Answer*  Slant height: If we slice the figure vertically, like a cake, the cross section is a trapezoid, as in Fig. 11–19. Since the trapezoid is regular, the triangles on either side are the same, so their base is 1 foot each: $\frac{(5 - 3)}{2}$.

Thus,

$$s^2 = 3^2 + 1^2$$
$$s = \sqrt{10}$$
$$s = 3.16 \text{ ft}$$

---

**Note:** *The same process can be used on all right frustums with an even number of sides (even conical ones).*

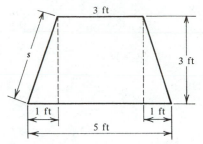

**Fig. 11–19**

$$\text{Average perimeter} = \frac{4(5\text{ ft}) + 4(3\text{ ft})}{2} = 16\text{ ft}$$

$$\text{Lateral surface area} = (3.16\text{ ft})(16\text{ ft})$$
$$= 50.56\text{ ft}^2$$

$$\text{Total surface area} = \text{LSA} + \text{Top area} + \text{Base area}$$
$$= 50.56 + 9 + 25$$
$$= 84.56\text{ ft}^2$$

### Practice Exercise No. 6

*(1)  Find the volume of Fig. 11–15 if the top and base are equilateral triangles (three equal sides) with sides of 7 cm and 10 cm, respectively, and an altitude of 5 cm.

 (2)  Find the volume and LSA of the frustum of a right pyramid with base 30 ft by 30 ft and top 20 ft by 20 ft, if the altitude is 25 ft.

*(3)  Find the volume and LSA of the frustum of a right cone with an altitude of 10 mm and diameters of 5 mm and 6 mm (top and bottom, respectively).

## 11.6  SPHERE

The volume of a sphere or ball is found by the formula

$$\text{Volume} = \frac{4\pi r^3}{3}$$

where $r$ is the radius of the ball. The surface area of the sphere is

$$\text{Surface area} = 4\pi r^2$$

---

**Note:**  *Volume uses $r^3$; surface area uses $r^2$.*

*Example (1)* Find the volume of a sphere with a 10″ diameter.

*Answer*

$$\text{Radius} = 5''$$

$$\text{Volume} = \frac{4(3.142)(5)^3}{3}$$

$$= \frac{4(3.142)(125)}{3}$$

$$= 523.7 \text{ in.}^3$$

*\*Example (2)* Find the surface area of a sphere with a diameter of 82 mm.

*Answer*

$$\text{Radius} = 41 \text{ mm}$$

$$\text{Surface area} = 4(3.142)(41)^2$$

$$= 4(3.142)(1681)$$

$$= 21\ 126.8 \text{ mm}^2$$

### Practice Exercise No. 7

Find the volume and surface area of the following figures.

(1) A sphere of radius = 3″.

\*(2) A sphere of radius = 30 cm.

(3) A sphere of diameter = 8 ft.

\*(4) A sphere of diameter = 77 mm.

## 11.7 PRACTICAL APPLICATIONS EXERCISE

(1) A swimming pool is 50 ft long, 20 ft wide, and is filled to a depth of 5 ft.
   (a) How many cubic feet of water are in the pool?
   (b) If 1 ft³ of water is equivalent to 7.48 gal, how many gallons of water are in the pool?

(2) The large standard size of concrete paving slab is 3′ × 2′ × 3″. If 1 ft³ of concrete has a mass of 125 lb, what is the mass of the slab?

\*(3) A copper tube is shown in Fig. 11–20. Calculate the volume of copper in the tube.

\*(4) A steel block is shown in Fig. 11–21.
   (a) Calculate its volume.
   (b) Calculate its mass if 1 cm³ of steel has a mass of 7.73 g.

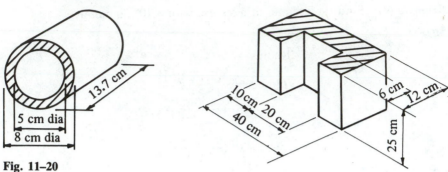

Fig. 11–20

Fig. 11–21

(5) A room measures 4.5 yd × 4.3 yd × 2.5 yd. Calculate how many cubic yards of air space are in the room.

(6) A circular container is to be designed to hold 11 000 in.³ of liquid. If the diameter of the container is to be 40″, calculate the height.

*(7) A trench is to be dug with the cross section as shown in Fig. 11–22; it is to be 55 m long. Calculate the volume of soil removed.

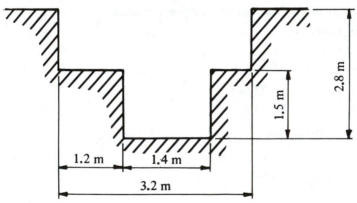

Fig. 11–22

*(8) An apprentice sketches two castings as shown in Figs. 11–23 and 11–24. If they are to be made of cast iron, which has a density of 7.28 g/cm³, calculate the mass of each casting.

(9) A steel ingot whose volume is 4 yd³ is rolled into a plate 1.5 yd wide and 10″ thick. Calculate the length of the plate in yards.

*(10) Water is flowing through a pipe 0.8 m in diameter at the rate of 1.5 m/sec. How many cubic meters of water will be delivered by the pipe in 2 min?

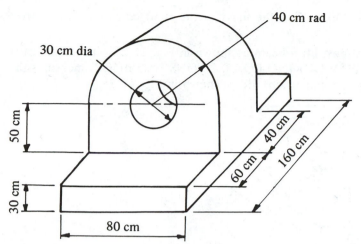

30 cm dia

40 cm rad

50 cm

30 cm

80 cm

60 cm

40 cm

160 cm

**Fig. 11–23**

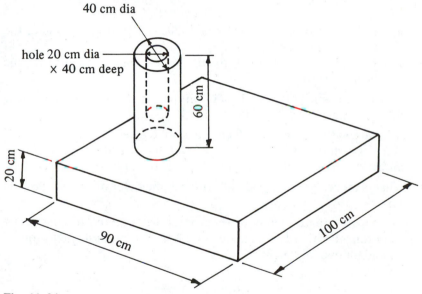

40 cm dia

hole 20 cm dia
× 40 cm deep

60 cm

20 cm

90 cm

100 cm

**Fig. 11–24**

*(11) A builder is asked to construct a circular fish pond of radius 3 m and depth 1.5 m. The pond is to have a path 1.3 m wide all round it.
(a) Calculate the volume of the pond in cubic meters.
(b) If 1 liter = 1000 cm³, calculate how many liters of water are required to fill the pond.
(c) Calculate the area of the path.

(12) A solid steel shaft is 8″ diameter and 18′ long. If 1 in.³ of steel has a

mass of 0.281 lb, find the mass of the shaft correct to the nearest pound.

*(13)  A worker must lift a beam as shown in Fig. 11–25 with a crane that says "SAFE WORKING LOAD 2 TONS." He roughly measures the beam and the dimensions are as shown in Fig. 11–25.

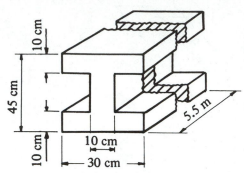

**Fig. 11–25**

(a)  If 1 m³ of beam material has a mass of 7830 kg, calculate the mass of the beam.

(b)  If 1 ton = 1016 kg, will the crane be strong enough to lift the beam?

(14)  Find the capacity of a steel pail, with a top diameter of 18″ and a bottom diameter of 16″, if the pail is 10″ deep.

*(15)  A square vat lid is made of 5 cm thick solid steel with a mass of 7.86 g/cm³. The top of the lid is 100 cm by 100 cm, and the bottom is 96 cm by 96 cm, forming a frustum. What is the mass of the lid?

(16)  What is the weight of a lead ball with a 3″ diameter if lead weighs 0.41 lb/in.³?

*(17)  How many square meters of sheet steel are required to construct a grain funnel with a top diameter of 3.1 m, a bottom diameter of 0.7 m, and an overall height of 2.1 m?

## 11.8   MISCELLANEOUS EXERCISES

*(1)  A casting 200 mm long is to be produced having a constant cross-sectional area as shown in Fig. 11–26 (dimensions in mm).

(a)  Determine the volume of the casting in cubic millimeters.

(b)  If the casting is produced from a circular ingot 40 mm dia × 160 mm long, determine the percentage of material used.

(2)  Find the volume of a piece of round brass rod 0.4″ in diameter and 42″ long.

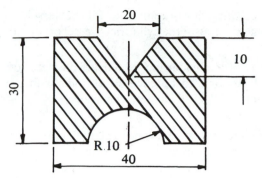

Fig. 11–26

*(3) Calculate the volume of the wedge shown in Fig. 11–27.

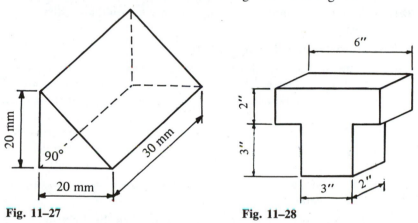

Fig. 11–27                                    Fig. 11–28

(4) Find the volume of the Tee section shown in Fig. 11–28.

*(5) Determine the volume of the wedge shown in Fig. 11–29.

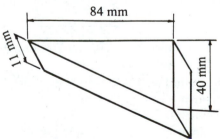

Fig. 11–29

(6) A solid copper cube, with sides of 45″, is placed inside an open cylindrical vessel with a diameter of 60″ and a height of 80″.

   **(a)**   Calculate the volume of water now required to fill the vessel.
   **(b)**   Calculate the percentage volume of the cylindrical vessel occupied by water.

 **(7)**   A hot water tank in the form of a cylinder has an outside diameter of 0.4 yd and a length of 1.2 yd. It is to be covered all over with an insulating material to a depth of 1.2″. Calculate the volume of insulating material required.

**\*(8)**   **(a)**   Calculate the capacity in *liters* of a cylindrical fuel tank with internal dimensions of 1 m diameter by 1.4 m long.

---

**Note:**   $\pi = \dfrac{22}{7}$   and   *1000 liters = 1 m³.*

---

   **(b)**   What length of tank of the same diameter would give 50% more capacity?

 **\*(9)**   Express 0.35 cm³ in cubic millimeters.

**\*(10)**   The cross section of a steel drill rod takes the form of a regular hexagon. The length of each side of the hexagon is 15 mm and the width across the flats may be taken to be 26 mm. The length of the rod is 2 m.
   **(a)**   Calculate the area of the cross section in square millimeters.
   **(b)**   Calculate the volume of steel in the rod in cubic centimeters.
   **(c)**   Calculate the area of the sides of the rod in square centimeters.

 **(11)**   Calculate the volume of concrete required for a garage base 8.000 yd long, 2.500 yd wide, and 0.150 yd thick.

**\*(12)**   **(a)**   A rectangular block of brass, measuring 5 cm by 3 cm by 2 cm, has a mass of 252 g. Calculate the density of brass.

---

**Note:**   *Density* $= \dfrac{Mass}{Volume}$.

---

   **(b)**   If the mass of the block is to be reduced to 210 g by drilling a hole in it, what volume of metal must be removed?

**\*(13)**   The special square-headed pin shown in Fig. 11–30 is to be produced from a 35 mm length of a 10-mm square cold-rolled bar.
   **(a)**   Calculate the volume of the pin.
   **(b)**   Calculate the percentage of the original bar machined away as scrap.

 **(14)**   A garage base, 18 ft by 9 ft, is to be filled with concrete to a depth of 6″. What will be the volume required? Give your answer in cubic yards.

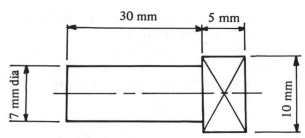

**Fig. 11–30**

**(15)** Find the volume of a cylinder 20″ in diameter by 10″ long.

**\*(16)** Figure 11–31 shows a hexagonal iron bolt with a cylindrical shaft of radius 3.5 mm. The broken line in figure A is the axis of rotational symmetry of the bolt.

  **(a)** Find the length $OX$.

  **(b)** Calculate the area of the hexagonal face in figure B.

  **(c)** Calculate the volume of the hexagonal head.

  **(d)** Calculate the volume of the cylindrical shaft (take $\pi \approx \dfrac{22}{7}$).

  **(e)** If the mass of 1 cm³ of iron is 8.5 g, find the mass of the bolt.

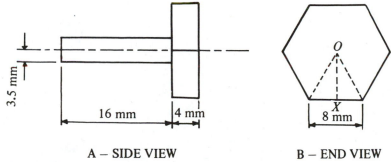

A – SIDE VIEW          B – END VIEW

**Fig. 11–31**

**\*(17)** Plastic rainwater guttering is made in rectangular sections (TYPE A) or semicircular sections (TYPE B). Both types are made in 2 m lengths with the other dimensions as shown in Figs. 11–32 and 11–33.

  **(a)** **(i)** Find the area of the flat piece of plastic needed to make a section of Type A guttering.

  **(ii)** Find the area of the flat piece of plastic needed to make a section of Type B guttering ($\pi \approx \dfrac{22}{7}$).

**(b)**    The ends of each section were blocked up and the gutters were filled with water. Calculate the volume of water in
   **(i)**    Type A.
   **(ii)**    Type B.

**(c)**    Which type holds the greater amount of water per square meter of plastic used?

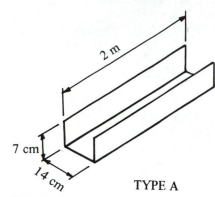

TYPE A

**Fig. 11–32**

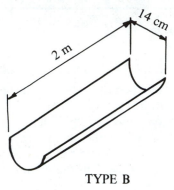

TYPE B

**Fig. 11–33**

# 12

# Geometrical Constructions

**Self-Assessment Test**

(1) Without using a protractor, construct an angle of 60°.

*(2) Construct a triangle with sides of 45 mm, 50 mm, and 60 mm. Draw its inscribed and circumscribed circles.

(3) Draw a line $3\frac{3''}{4}$ long. Divide the line into 5 equal parts by construction; that is, do not use a rule for division.

*(4) A rectangle 100 mm long by 70 mm wide is to have its corners shaped into circular arcs of radius 20 mm. Draw the resulting outline showing all construction lines (especially how to obtain centers for corner radii).

*(5) Draw the following shape showing all construction lines.

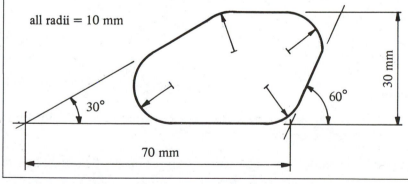

## 12.1 BASIC CONSTRUCTIONS

Craftsmen often require the ability to mark out accurately a component in readiness for manufacture. In many instances, they will use geometrical equipment to construct shapes. It is therefore essential that they have an understanding of basic geometrical constructions.

**(1)   To divide a line into two equal parts (or to bisect a line).**

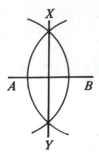

Method:

**(a)**   With $A$ and $B$ as centers and a radius bigger than $\frac{1}{2}AB$, draw arcs that intersect at $X$ and $Y$.

**(b)**   Join $XY$. The line $XY$ will divide $AB$ into two equal parts and will be perpendicular (at right angles) to $AB$.

**(2)   To bisect an angle.**

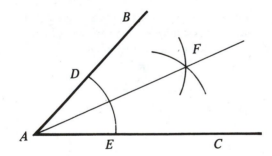

Method:

**(a)**   With center $A$ and any radius, draw an arc to cut $AB$ and $AC$ at $D$ and $E$, respectively.

**(b)**   With centers $D$ and $E$ and radius bigger than $\frac{1}{2}DE$, draw arcs to intersect at $F$.

**(c)**   Join $A$ to $F$. The line $AF$ will bisect angle $BAC$.

**(3)   To construct a perpendicular line from a given point A on a straight line.**

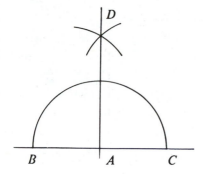

Method:

**(a)** With center $A$ and any radius, draw an arc to cut the straight line at $B$ and $C$.

**(b)** With centers $B$ and $C$ and a radius greater than $AB$, draw arcs to intersect at $D$.

**(c)** Join $A$ and $D$. The line $AD$ will be perpendicular to $BC$.

**(4)** **To construct a perpendicular to a given line from a point A that is not on the line.**

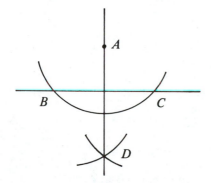

Method:

**(a)** With $A$ as center, draw an arc to cut the given straight line at $B$ and $C$.

**(b)** With $B$ and $C$ as centers and a radius bigger than $\dfrac{1}{2}BC$, draw two arcs to intersect at $D$.

**(c)** Join $AD$. $AD$ will be perpendicular to the given line.

**(5)** **To construct a line parallel to a given line at a given distance.**

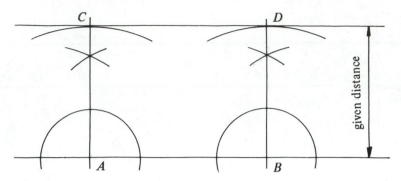

Method:

**(a)** Choose two points, $A$ and $B$, on the given straight line.

**(b)** Construct perpendicular lines at $A$ and $B$ (see point 3).

**(c)** With $A$ and $B$ as centers, mark off the given distance along the perpendicular lines to cut at $C$ and $D$.

**(d)** Join $C$ and $D$ which will produce the required line.

**(6)   To construct an angle of 60°.**

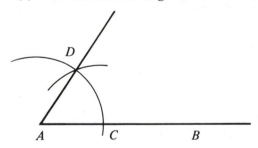

Method:

**(a)** Draw a straight line $AB$.

**(b)** Center $A$ and any radius; draw an arc to cut $AB$ at $C$.

**(c)** With $C$ as center and using the same radius, draw another arc to cut the first one at $D$.

**(d)** Join $DA$. Angle $DAB$ will be 60°.

**(7)   To construct an angle of 30°.**

Method:

**(a)** Construct an angle of 60° (see point 6).

**(b)** Bisect it (see point 2) to create a line at 30°.

**(8)** **To construct an angle of 45°.**

Method:

**(a)** Construct a right angle (see point 3).

**(b)** Bisect it (see point 2) to create a line at 45°.

**(9)** **To divide a line, AB, into a given number of equal parts (e.g., 6 parts).**

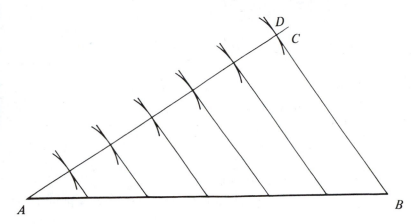

Method:

**(a)** Draw $AC$ at any angle to $AB$.

**(b)** Using compasses, mark off 6 equal divisions along $AC$, starting from $A$.

**(c)** Join the last mark $D$ to $B$.

**(d)** From each of the other marks on $AC$, construct lines parallel to $DB$ to cut $AB$. $AB$ will then be divided into 6 equal parts.

**(10)** **To construct a triangle given the lengths of the sides.**

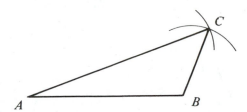

Method:

**(a)** Draw a line $AB$ equal in length to one of the sides.

**(b)** Center $A$; draw an arc whose radius is equal to another side.

(c)   Center $B$; draw an arc whose radius is equal to the third side to cut the previous arc at $C$.

(d)   Join $AC$ and $BC$ to form the required triangle.

**(11)   To find the center of a circle or an arc.**

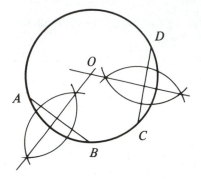

Method:

(a)   Draw any two chords, $AB$ and $CD$.

(b)   Draw the perpendicular bisectors of these chords (see point 1).

(c)   Where the perpendicular bisectors of the chords cross, $O$ will be the center.

**(12)   To construct the circumscribed circle of a given triangle.**

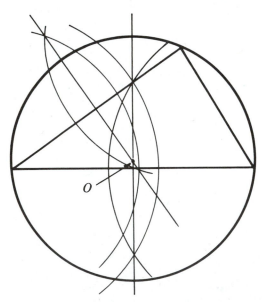

Method:

**(a)** Construct the perpendicular bisectors of two sides (see point 1).

**(b)** Where these lines intersect, $O$ will give the center of the circumscribed circle.

**(13)** **To construct the inscribed circle of a given triangle.**

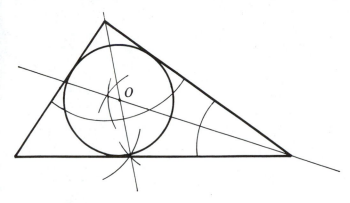

Method:

**(a)** Construct the bisectors of two internal angles (see point 2).

**(b)** Where these lines intersect, $O$ will give the center of the inscribed circle.

**(14)** **To construct a circle of given radius to touch two given straight lines.**

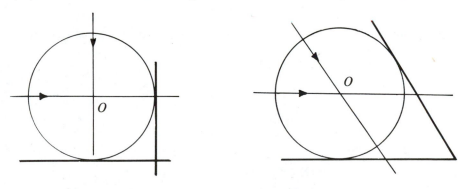

Method:

**(a)** Draw two lines, each parallel to one of the given lines at a distance equal to the radius of the circle from it.

**(b)** Where these lines intersect, $O$ will give the center of the required circle.

**(15)** **To construct a circle of given radius to touch a straight line and a given circle.**

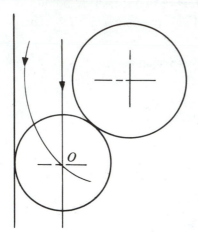

Method:

**(a)** Draw a line parallel to the given straight line at a distance equal to the radius of the required circle.

**(b)** Add the radius of the required circle to the radius of the given circle and draw an arc whose center is at the center of the given circle to intersect the previous construction line at *O*. *O* is the center of the required circle.

**(16)** **To draw a circle or arc of a given radius to touch two given circles.**

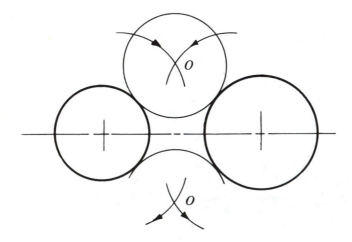

Method:

(a) Add the radius of the required circle or arc to the radii of both circles.

(b) Using the centers of the existing circles, draw arcs to intersect at *O*.

(c) *O* is the center of the required circle or arc.

### Practice Exercise

*(1) Draw a line 95 mm long and bisect it.

(2) Construct an angle of 60°. From this construct an angle of 15°.

(3) Draw a line 7″ long and, by construction, divide it into 3 equal parts.

(4) Draw a line $3\frac{3''}{8}$ long and, by construction, divide it into 7 equal parts.

(5) Construct a triangle with sides 6″, 7″, and 8″.

*(6) Construct a triangle with sides 35 mm, 55 mm, and 50 mm.

(7) Construct a triangle with sides $1\frac{1''}{2}$, $1\frac{5''}{8}$, and 2″ and then draw its inscribed and circumscribed circles.

*(8) A rectangle 90 mm long by 60 mm wide is to have its corners shaped into circular arcs of 15 mm radius. Draw the resulting outline showing all construction lines.

*(9) Draw a circle, 30 mm in diameter, to touch the circles below.

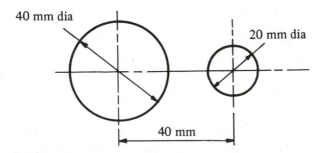

40 mm dia

20 mm dia

40 mm

## 12.2  PRACTICAL APPLICATIONS EXERCISE

(1) You are given a triangular piece of metal with sides 10″, 9″, and 8″. You are asked to drill a 3″ diameter hole in the component so that the center is equidistant from the three sides, as in Fig. 12–1. Construct

the triangle and, showing all construction lines, draw the circle representing the hole.

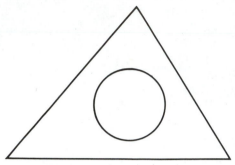

**Fig. 12–1**

(2) You are given a solid circular disc, 6″ in diameter as shown in Fig. 12–2. You are asked to insert a 1″ diameter hole in the center of the disc. Draw the disc and, showing all construction lines, show how the center hole could be marked out.

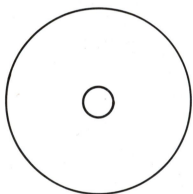

**Fig. 12–2**

(3) A decorative panel is to be made up of several sections, one of which is shown in Fig. 12–3. Draw, full size, the above shape showing all construction lines. Measure the diameters of the small and large circles.

(4) An artist designs a medallion as shown in Fig. 12–4. It is to be made completely from a long thin strip of metal and consists of an equilateral triangle with sides of 8″ and two circular pieces of material attached as shown. Construct the triangle and, showing all construction lines, the inner and outer circles. Give the diameters of the circles and calculate the total length of material required.

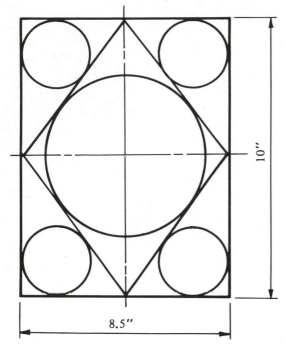

Fig. 12–3

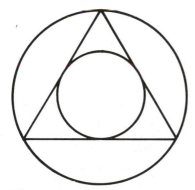

Fig. 12–4

*(5)  A thin plate is to be drawn and cut as shown in Fig. 12–5. Draw its outline, full size.

(6)  The part illustrated in Fig. 12–6 is to be constructed. An apprentice decides that she will first mark out a triangle with sides 7.5, 9.0, and 10.0″. She will then construct the centers for the curves and the circle (which is to be equidistant from the sides). Draw, full size, the component showing all construction lines.

10 mm rad

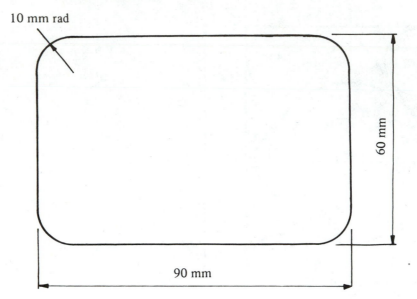

90 mm

60 mm

**Fig. 12–5**

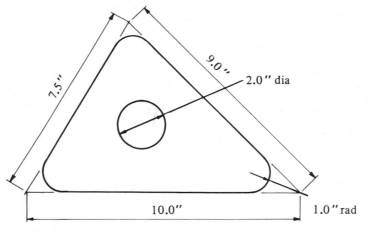

7.5″

9.0″

2.0″ dia

10.0″

1.0″ rad

**Fig. 12–6**

*(7) You are asked to produce the component in Fig. 12–7. Draw the shape, full size, showing all construction lines. It is symmetrical about the major centerlines.

*(8) You are requested to make a special wrench as shown in Fig. 12–8. Draw the shape, full size, showing all the construction lines. It is to be symmetrical about the centerline.

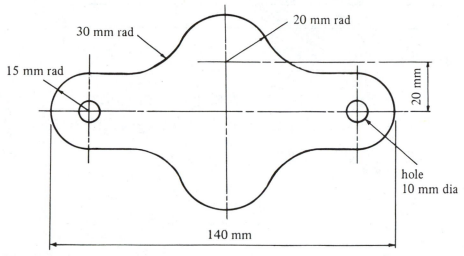

**Fig. 12–7**

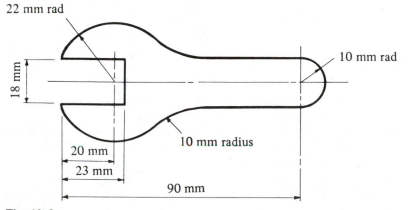

**Fig. 12–8**

**(9)** A man wishes to produce a drawing to a scale of $3\frac{1}{2}''$ to 1 ft. Draw a line $3\frac{1}{2}''$ long and divide it into 12 equal parts so that inches may be measured for the drawing. Label lengths of 3", 6", 9", and 12". Show all construction lines.

**\*(10)** A gauge has to be calibrated to read liters of water. The line shown in Fig. 12–9 appears on the gauge and, if the indicator reaches the end, this represents 10 liters. Divide the line into 10 equal parts and label 0, 2, 4, 6, 8, and 10 liters. Show all construction lines.

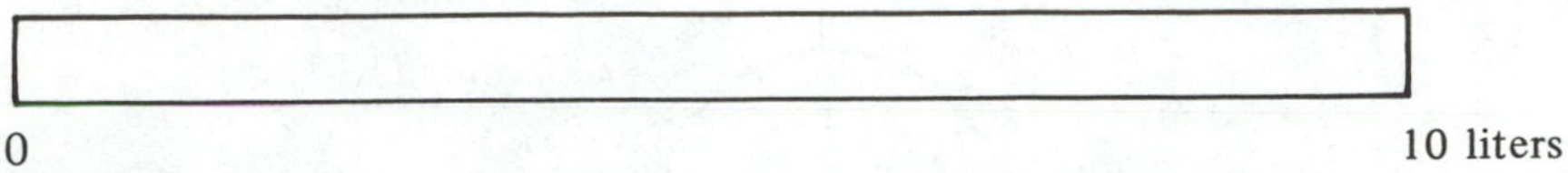

0                                                                        10 liters

**Fig. 12–9**

## 12.3   MISCELLANEOUS EXERCISES

**(1)**   Draw the circle in Fig. 12–10 and add the basic geometry that will locate its center point. Diameter $= 1\dfrac{5''}{8}$ .

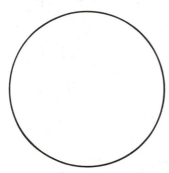

**Fig. 12–10**

**\*(2)**   A rectangular steel plate *ABCD* is 6 mm thick and has two adjacent faces *AB* (55 mm) and *BC* (75 mm) as datum edges. Three holes *X*, *Y*, and *Z* are to be drilled in the plate. The center of hole *X* (8 mm dia) is 18 mm from *AB* and 12 mm from *BC*. The center of hole *Y* (10 mm dia) is 42 mm from *AB* and on a line drawn from *B* at an angle of 45° to *BC*. The center of hole *Z* (12 mm dia) is 45 mm from the center of hole *X* and 30 mm from the center of hole *Y*. Draw the plate twice full size; fully dimension it using the centerline of hole *X* and the edge *BC* as datums. (You may measure the distance for hole *Z* from your drawing.)

**(3)**   Using only a pair of compasses and a ruler, divide a circle 7″ in diameter into six equal parts.

# PART IV

## Applied Trigonometry

# 13

# Angles

**Self-Assessment Test No. 1**

Add the following angles.

**(1)** 13°21′13″ and 16°30′10″  **(2)** 69°31′19″ and 45°45′50″

**(3)** 77°18′40″ and 17°37′12″  **(4)** 88°45′37″ and 16°41′27″

Subtract the following angles.

**(5)** 88°10′13″ from 99°50′28″  **(6)** 37°18′20″ from 51°17′55″

**(7)** 17°41′5″ from 20°13′4″  **(8)** 14°59′59″ from 100°

**(9)** Convert 87.51° to degrees/minutes/seconds.

**(10)** Convert 18°15′15″ to decimal degrees.

The term **angle** is used to denote an amount of turning or rotation. In Fig. 13–1, the line *OX* has been rotated about *O* in a **counterclockwise** direction until it takes up the position *OY*. The amount of turning is called angle *XOY*.

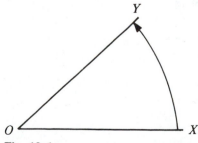

**Fig. 13–1**

Angles are usually measured in degrees, minutes, and seconds where

60 seconds = 1 minute     Notation: 60″ = 1′

221

60 minutes = 1 degree        Notation: 60′ = 1°

360 degrees = 1 revolution

Usually for general craft work, only degrees are stated. For very accurate work minutes and seconds are used.

## 13.1   ADDITION AND SUBTRACTION OF ANGLES

*Example (1)*   Add 13°51′ and 22°12′.

*Answer*

$$
\begin{array}{r}
13°51′ \\
+\ 22°12′ \\
\hline
36°3′
\end{array}
$$
Note that 51′ + 12′ = 63′ = 1°3′
(since 60′ = 1°)

*Example (2)*   Subtract 18°25′33″ from 37°44′18″.

*Answer*

$$
\begin{array}{r}
37°44′18″ \\
-\ 18°25′33″ \\
\hline
19°18′45″
\end{array}
$$

## Practice Exercise No. 1

Add the following angles.

**(1)**   25°8′ and 16°37′

**(2)**   14°57′ and 29°31′

**(3)**   87°14′ and 81°59′

**(4)**   9°38′37″ and 47°31′49″

**(5)**   81°59′40″ and 6°13′57″

**(6)**   63°22′31″ and 12°13′41″

**(7)**   85°37′19″ and 45°18′55″

**(8)**   77°19′16″ and 16°43′50″

**(9)**   1°5′55″ and 2°49′51″

**(10)**   13°13′40″ and 45°59′48″

Subtract the following angles.

**(11)**   7°9′ from 22°18′

**(12)**   17°37′ from 22°13′

**(13)**   16°38′22″ from 31°40′50″

**(14)**   88°16′32″ from 110°8′15″

**(15)**   103°17′44″ from 110°

**(16)**   71°18′19″ from 91°6′32″

## 13.2   CONVERSIONS BETWEEN DECIMAL DEGREES AND DEGREES/MINUTES/SECONDS

Frequently a problem will require an angle to be converted from decimal degrees to degrees/minutes/seconds, or vice versa. In the trigonometry chapters we will work principally with decimal degrees, so that we will have to convert to degrees, minutes, and seconds.

## (1) Decimal Degrees to Degrees/Minutes

Multiply the decimal portion of the number by 60 minutes (60′).

*Example (1)* Convert 18.7° to degrees/minutes.

$$18.7° = 18° \underset{\quad\quad\;\searrow\; 0.7 \times 60' = 42'}{?\,'}$$

Therefore,

$$18.7° = 18°42'$$

## (2) Decimal Degrees to Degrees/Minutes/Seconds

(a) Convert to degrees/minutes.

(b) Multiply the decimal portion of the minutes by 60 seconds (60″).

*Example (2)* Convert 65.94° to degrees/minutes/seconds.

$$65.94° = 65° \underset{\searrow\; 0.94 \times 60' = 56.4'}{?\,'}$$
$$= 65°56.4'$$
$$= 65°56' \underset{\searrow\; 0.4 \times 60'' = 24''}{?\,''}$$
$$= 65°56'24''$$

## (3) Converting Degrees/Minutes/Seconds to Decimal Degrees

(a) Convert the seconds to minutes by dividing by 60″ and add to the previous minutes.

(b) Convert the minutes to degrees by dividing by 60′ and add to the previous degrees.

*Example (3)* Convert 187°44′15″ to decimal degrees.

$$15 \div 60 = 0.25'$$
$$\therefore 187°44'15'' = 187°44.25'$$
$$44.25' \div 60 = 0.7375°$$
$$\therefore 187°44'15'' = 187.7375°$$

## Practice Exercise No. 2

Convert to degrees/minutes/seconds.

**(1)** 17.51°

**(2)** 87.913°

**(3)** 47.3°

**(4)** 147.843°

Convert to decimal degrees.

**(5)** 22°18′

**(6)** 45°45′

**(7)** 88°8′15″

**(8)** 91°6′32″

---

*Self-Assessment Test No. 2*

**(1)** Define an acute angle.

**(2)** Define a reflex angle.

**(3)** What is an equilateral triangle?

Calculate the lettered angles in the following figures.

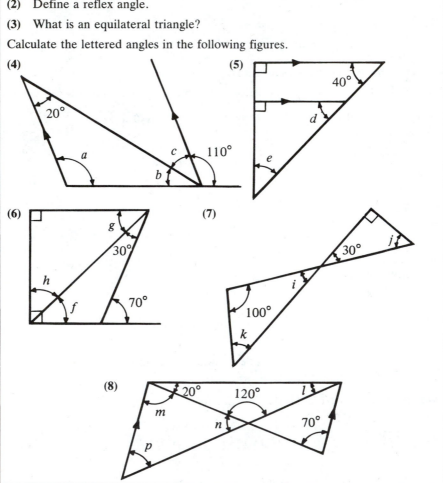

## 13.3 MEASUREMENT OF ANGLES

If angles involving minutes and seconds are to be measured, very accurate measuring devices are required. If the degree of accuracy is not high, then angles that are given just in degrees may be measured using a *protractor*, as shown in Fig. 13–2.

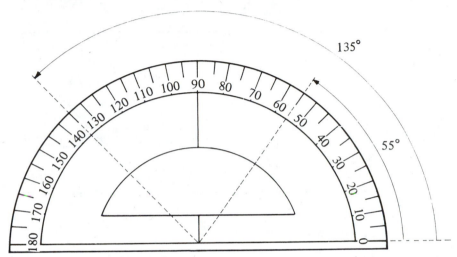

**Fig. 13–2**

### Practice Exercise No. 3

Trace the following angles, measure them using a protractor, and label (as in Fig. 13–3).

> **Hint:** *You may have to make the sides of the angle longer to read the angle on the protractor.*

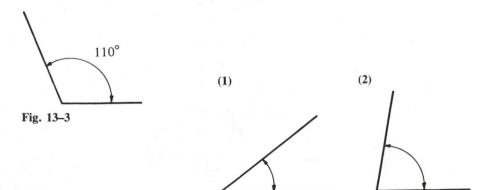

110°

**(1)**          **(2)**

**Fig. 13–3**

(3)                     (4)                     (5)

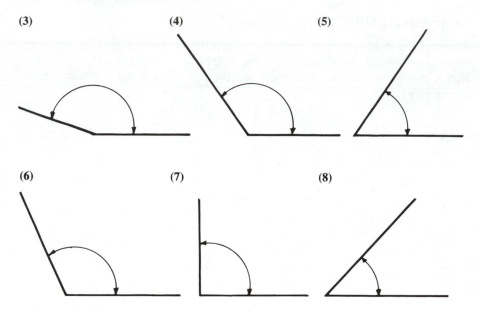

(6)                     (7)                     (8)

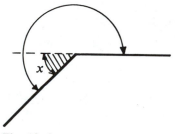

## For Angles Bigger than 180°

***Example (1)***   In Fig. 13–4,

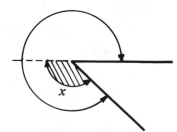

**Fig. 13–4**                    **Fig. 13–5**

$$\text{Angle} = 180° + 45°$$
$$= 225°$$

***Example (2)***   In Fig. 13–5,

$$\text{Angle} = 180° + 135°$$
$$= 315°$$

Measure the small angle $x$, and then add to 180° to obtain the required angle.

## Practice Exercise No. 4

Trace the following angles, measure them using a protractor, and label (as in Fig. 13–6).

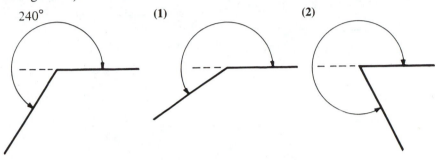

240°   (1)   (2)

Fig. 13–6

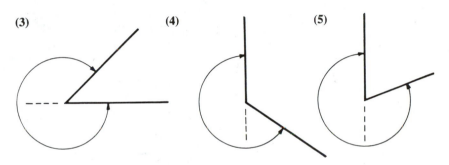

(3)   (4)   (5)

## Practice Exercise No. 5

Use a protractor to draw the following angles.

| | | | |
|---|---|---|---|
| **(1)** 45° | **(2)** 85° | **(3)** 15° | **(4)** 154° |
| **(5)** 260° | **(6)** 330° | **(7)** 270° | **(8)** 71° |
| **(9)** 340° | **(10)** 33° | **(11)** 105° | **(12)** 205° |

## 13.4   TYPES OF ANGLES

### Acute Angles

An acute angle is less than 90°. See Fig. 13–7.

### Obtuse Angles

An obtuse angle is greater than 90° but less than 180°. See Fig. 13–8.

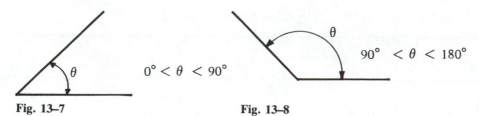

$$0° < \theta < 90°$$

**Fig. 13–7**

$$90° < \theta < 180°$$

**Fig. 13–8**

### Reflex Angles

A reflex angle is greater than 180° but less than 360°. See Fig. 13–9.

### Right Angles

A right angle is equal to 90°. See Fig. 13–10.

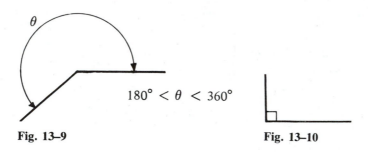

$$180° < \theta < 360°$$

**Fig. 13–9**

**Fig. 13–10**

### Complementary Angles

These are two angles whose sum is 90°. See Fig. 13–11.

### Supplementary Angles

These are two angles whose sum is 180°. See Fig. 13–12.

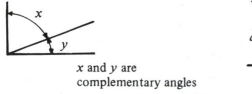

$x$ and $y$ are
complementary angles

**Fig. 13–11**

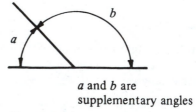

$a$ and $b$ are
supplementary angles

**Fig. 13–12**

## 13.5  PROPERTIES OF
## ANGLES AND STRAIGHT LINES

**(1)** The total angle on a straight line is equal to 180°. For example, see Fig. 13–13.

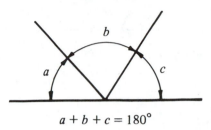

$$a + b + c = 180°$$

**Fig. 13–13**

**(2)** When two straight lines cross, opposite angles are equal. For example, see Fig. 13–14.

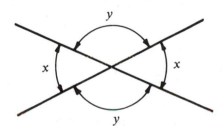

**Fig. 13–14**

**(3)** If a line cuts two parallel lines, opposite and corresponding angles are equal. For example, in Fig. 13–15, corresponding angles are equal and, in Fig. 13–16, opposite angles are equal.

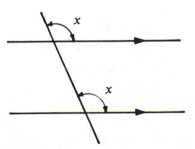

**Fig. 13–15**

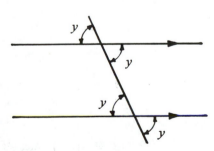

**Fig. 13–16**

**(4)** If a line cuts two parallel lines alternate angles are equal. For example, in Figs. 13–17 and 13–18, alternate angles are equal.

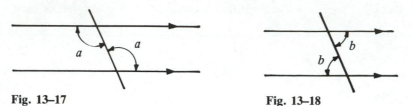

**Fig. 13–17**                    **Fig. 13–18**

## 13.6 ANGLE PROPERTIES OF TRIANGLES

**(1)** The sum of the angles in a triangle is 180°. See Fig. 13–19.

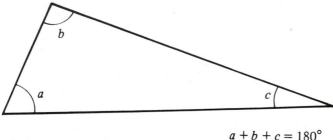

$$a + b + c = 180°$$

**Fig. 13–19**

**(2)** A **scalene** triangle has all three sides of different length and all angles are different. See Fig. 13–20.

**(3)** An **isosceles** triangle has two sides and two angles equal. See Fig. 13–21.

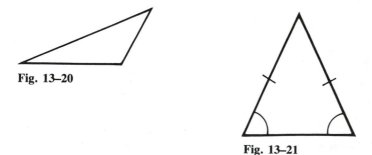

**Fig. 13–20**

**Fig. 13–21**

**49**
**50**

**(4)** An **equilateral** triangle has all its sides equal and all angles equal. Each angle of the triangle is 60°. See Fig. 13–22.

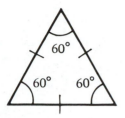

Fig. 13–22

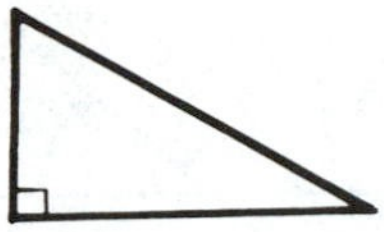

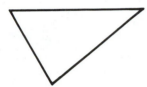

Fig. 13–23

**(5)** A **right** triangle contains a right angle and two acute angles. See Fig. 13–23.

**(6)** An **acute** triangle has three acute angles. See Fig. 13–24.

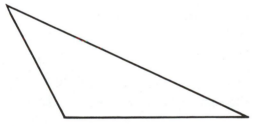

Fig. 13–24

**(7)** An **obtuse** triangle has one obtuse angle. See Fig. 13–25.

All of this information is very useful when attempting to do calculations involving angles.

Fig. 13–25

***Example (1)*** Calculate angles $r$, $s$, and $t$ in Fig. 13–26.

$s = 50°$     When two straight lines cross, opposite angles are equal.

$\left. \begin{array}{l} r + s = 180° \\ r + 50° = 180° \end{array} \right\}$   Total angles on a straight line $= 180°$.

$r = 130°$

$t = 50°$     $t$ is corresponding to $s$; corresponding angles equal.
                Also, $t$ is alternate to $50°$, alternate angles equal.

***Answer***  $r = 130°$, $s = 50°$, $t = 50°$.

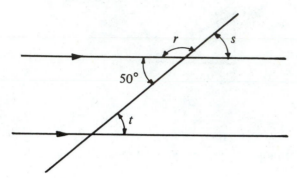

**Fig. 13–26**

***Example (2)***   Calculate angles *x, y,* and *z* in Fig. 13–27.

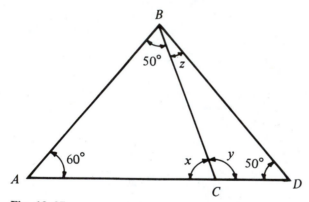

**Fig. 13–27**

Consider the triangle *ABC*. The sum of all angles inside a triangle = 180°.

$$\therefore 60° + 50° + x = 180°$$
$$110° + x = 180°$$
$$x = 70°$$

If *x* = 70°, *y* = 180° − 70° = 110°. Now consider the triangle *BCD*. The sum of all angles inside a triangle = 180°.

$$\therefore 110° + 50° + z = 180°$$
$$160° + z = 180°$$
$$z = 20°$$

***Answer***   *x* = 70°, *y* = 110°, *z* = 20°.

## Practice Exercise No. 6

Calculate the lettered angles in the following figures.

**(1)**

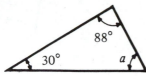

**(2)**

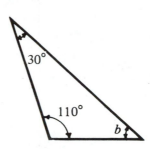

**(3)**

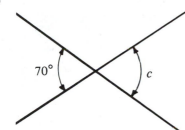

**(4)**

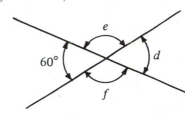

**(5)**

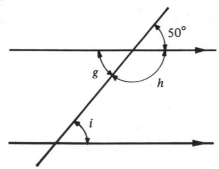

**(6)**

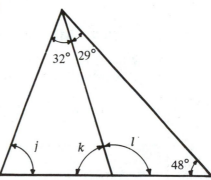

(7) (8)

(9) (10)

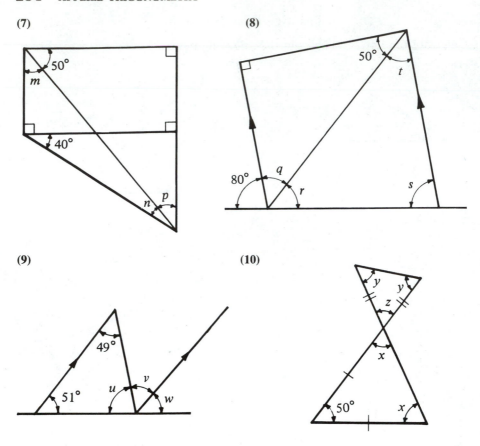

## 13.7 MISCELLANEOUS EXERCISES

*(1) (a) What is the angle θ in Fig. 13–28?
(b) Calculate the area of the shape in Fig. 13–28.

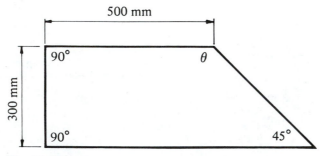

Fig. 13–28

(2) In Fig. 13–29, *AB* is parallel to *CD* and *XY* is perpendicular to *AB*. Calculate the angles marked *p* and *q*.

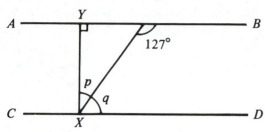

**Fig. 13–29**

(3) In triangle *PQR*, angle *PQR* = 54° and angle *PRQ* = 62°.
   (a) Calculate angle *RPQ*.
   (b) Which is the longest side of triangle *PQR*?

(4) Figure 13–30 shows a square with an isosceles triangle subtended on each side.

This diagram is not drawn to scale

Angle *EBF* = 130°

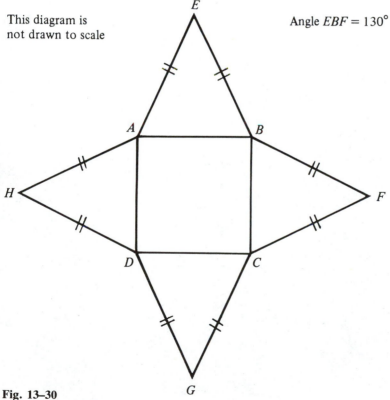

**Fig. 13–30**

    **(a)**  Calculate the size of angle *FBC*.

    **(b)**  Calculate the size of angle *BFC*.

**(5)**  If the smallest angle of a triangle is 23° and the difference between the other two angles is 39°, calculate the largest angle of the triangle.

# 14

# Trigonometry of Right Triangles

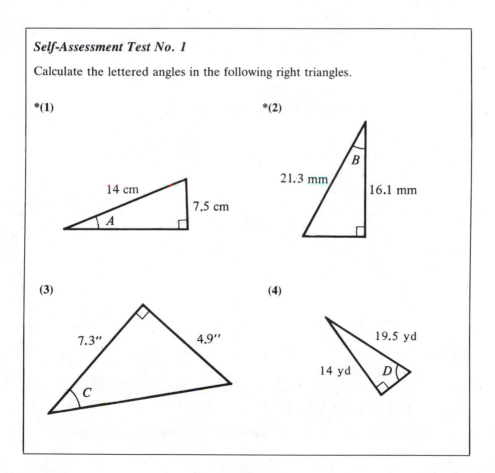

*Self-Assessment Test No. 1*

Calculate the lettered angles in the following right triangles.

*(1)

14 cm

7.5 cm

*A*

*(2)

21.3 mm

*B*

16.1 mm

(3)

7.3"

4.9"

*C*

(4)

19.5 yd

14 yd

*D*

## 14.1   DEFINITIONS

Angles are usually measured in degrees. Degrees can be split into smaller units called minutes where 1 degree = 60 minutes (or 60′). For example, 30°30′ is equal to $30\frac{1}{2}$ degrees.

Consider the right triangle shown in Fig. 14–1. The names of the sides with respect to angle $A$ are:

Hypotenuse: Longest side and always opposite to the right angle.

Adjacent: Side nearest to angle $A$ other than hypotenuse.

Opposite: Side opposite to angle $A$.

(Notice that these relationships would change if we used the other angle.)

In a right triangle, the ratios of the sides are called trigonometric ratios. These ratios are:

$$\text{Sine of angle } A, \text{ or } \sin A = \frac{\text{opposite}}{\text{hypotenuse}}$$

$$\text{Cosine of angle } A, \text{ or } \cos A = \frac{\text{adjacent}}{\text{hypotenuse}}$$

$$\text{Tangent of angle } A, \text{ or } \tan A = \frac{\text{opposite}}{\text{adjacent}}$$

These ratios, together with sine, cosine, and tangent tables, are very useful when finding angles and lengths of sides in **right triangles**.

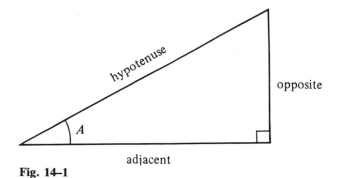

**Fig. 14–1**

## 14.2   USING THE TABLES

The values of the ratios at various angles are contained in Table D2 of Appendix D.

## Finding Values

Table D2 is simple to use if we are looking up the value of a ratio when we know the angle.

**(1)** If the angle is less than 45°, find the angle in the *left* column and the ratio at the *top* of the page. Where they meet is the value of the ratio.

> ***Example*** sin 35.4° = 0.5793
>
> cos 26.7° = 0.8934
>
> tan 43.8° = 0.9590

**(2)** If the angle is between 45° and 90°, find the angle in the *right* column and the ratio at the *bottom* of the page (notice that the ratios are not the same at the top and the bottom). Where they meet is the value of the ratio.

> ***Example*** sin 47.9° = 0.7420
>
> cos 73.4° = 0.2857
>
> tan 87.5° = 22.90

## Practice Exercise No. 1

Neatly copy and complete the following table.

| Angle | Sine | Cosine | Tangent |
|-------|------|--------|---------|
| 50.7° | 0.7738 | | |
| 30° | | | |
| 60° | | | |
| 10.2° | | | |
| 18.9° | | | |
| 82.9° | | 0.1236 | |
| 21.2° | | | |
| 33.7° | | | 0.6669 |

## Finding Angles

If we know the value of a ratio and need to find the angle, we need to use the same table in reverse.

*Example* Find the angle whose sine is 0.1719.

*Answer* 9.9°

*Example* Find the angle whose sine is 0.8721.

*Answer* 60.7°

Notice that here we found the answer in the column with *sin* at the bottom, so we found the angle in the right column.

*Example* Find $\theta$ if sin $\theta$ = 0.7815.

*Answer* 51.4°

*Example* Find $\theta$ if tan $\theta$ = 1.422.

*Answer* 54.9°

*Example* Find *A* if cos *A* = 0.8342.

*Answer* 33.5°

Notice that, if we could not find the exact value in the body of the table, we took the closest value to ours to find the angle.

## Practice Exercise No. 2

| Angle | Sine | Cosine | Tangent |
|---|---|---|---|
| | 0.6252 | | |
| | | 0.4274 | |
| | | | 2.9544 |
| 41.8° | | | |
| | 0.8039 | | |
| | | | 7.806 |
| | | 0.9976 | |
| 87.4° | | | |
| | 0.0645 | | |

---

**Calculator Hint**

If you are using a calculator with this chapter, you should refer to Appendix C, Section C.4.

## 14.3 USING TRIGONOMETRY TO FIND ANGLES IN A RIGHT TRIANGLE

*\*Example (1)*  Find angle $X$.

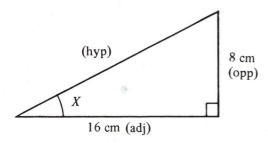

Labeling the sides, the sides known are **opp** and **adj**. Which trigonometric ratio can we use—sine, cosine, or tangent?

*Answer*  Tangent

$$\tan X = \frac{\text{opp}}{\text{adj}} = \frac{8}{16} = \frac{1}{2} = 0.5000$$

By looking up 0.5000 in tangent tables, we find that

$$\text{angle } X = 26.6°$$

*Example (2)*  Find angle $Y$.

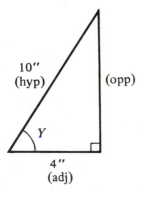

The sides known are **adj** and **hyp**. Which trigonometric ratio uses adj and hyp?

*Answer*  Cosine

$$\cos Y = \frac{\text{adj}}{\text{hyp}} = \frac{4}{10} = 0.4000$$

By looking up 0.4000 in cosine tables, we find that

$$\text{angle } Y = 66.4° \quad \text{or} \quad 66°24' \quad (\text{see Section 13.2})$$

*Example (3)*   Find angle $Z$.

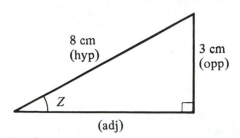

The sides known are **opp** and **hyp**. Which trigonometric ratio uses opp and hyp?

*Answer*   Sine

$$\sin Z = \frac{\text{opp}}{\text{hyp}} = \frac{3}{8} = 0.3750$$

By looking up 0.3750 in sine tables, we find that

$$\text{angle } Z = 22.0°$$

*Example (4)*   Find angle $X$.

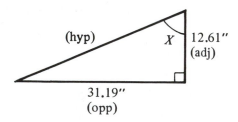

$$\tan X = \frac{\text{opp}}{\text{adj}} = \frac{31.19}{12.61} = 2.474$$

$$\text{angle } X = 68.0°$$

## Practice Exercise No. 3

Calculate the lettered angles in the following triangles.

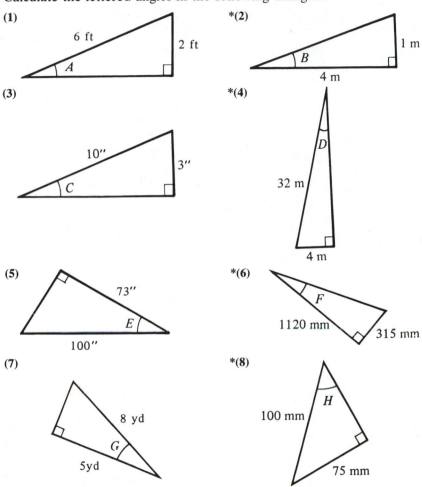

**(1)**

6 ft

2 ft

*A*

**\*(2)**

1 m

*B*

4 m

**(3)**

10″

3″

*C*

**\*(4)**

*D*

32 m

4 m

**(5)**

73″

*E*

100″

**\*(6)**

*F*

1120 mm

315 mm

**(7)**

8 yd

*G*

5yd

**\*(8)**

*H*

100 mm

75 mm

## Practice Exercise No. 4

Calculate the lettered angles in the following triangles.

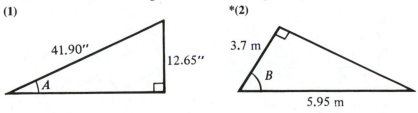

**(1)**

41.90″

12.65″

*A*

**\*(2)**

3.7 m

*B*

5.95 m

(3)

*(4)

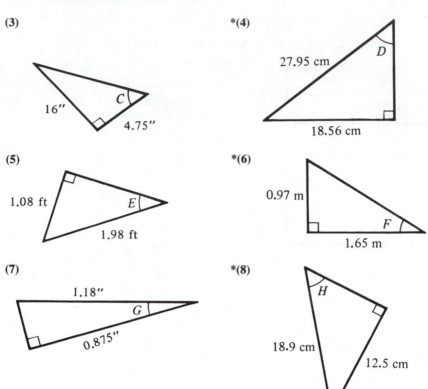

16"

C

4.75"

27.95 cm

D

18.56 cm

(5)

*(6)

1.08 ft

E

1.98 ft

0.97 m

F

1.65 m

(7)

*(8)

1.18"

G

0.875"

H

18.9 cm

12.5 cm

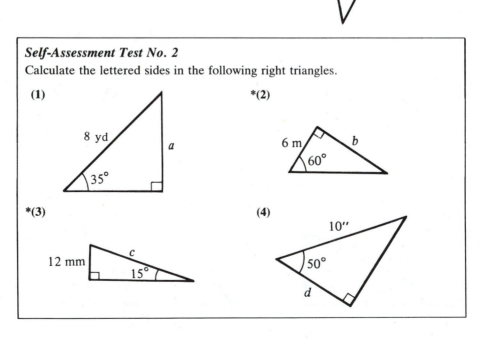

---

## Self-Assessment Test No. 2

Calculate the lettered sides in the following right triangles.

**(1)**

8 yd

a

35°

*(2)

6 m

b

60°

*(3)

12 mm

c

15°

(4)

10"

50°

d

## 14.4  USING TRIGONOMETRY TO FIND LENGTHS OF SIDES IN A RIGHT TRIANGLE

*\*Example (1)*  We know that a particular angle is 35° and the hypotenuse is 8 cm. We wish to find the opposite side (opp) or $x$. What trigonometric ratio connects the angle, hyp, and opp?

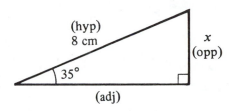

*Answer*  Sine

$$\sin 35° = \frac{\text{opp}}{\text{hyp}} = \frac{x}{8}$$

Multiply both sides by 8,

$$8 \sin 35° = \frac{x}{\cancel{8}} \times \cancel{8}$$

Thus,

$$x = 8 \sin 35°$$
$$= 8 \times 0.5736$$
$$= 4.5888 \text{ cm}$$

*Example (2)*  What trigonometric ratio connects the angle, opp, and adj in the following figure.

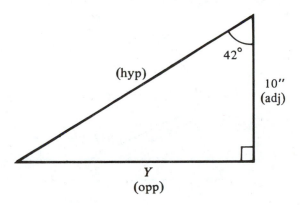

*Answer*   Tangent

$$\tan 42° = \frac{\text{opp}}{\text{adj}}$$

$$= \frac{Y}{10}$$

Multiply both sides by 10,

$$10 \tan 42° = \frac{Y}{10} \times 10$$

$$Y = 10 \times 0.9004$$

$$= 9.004''$$

## Practice Exercise No. 5

Calculate the lettered sides in the following right triangles.

**\*(1)**

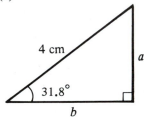

**(2)**

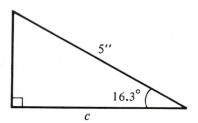

**\*(3)**

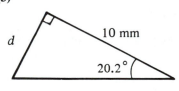

**(4)**

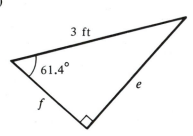

**\*(5)**

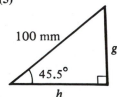

**(6)**

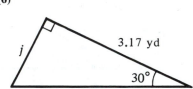

*(7)

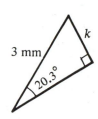

(8)

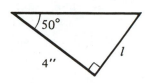

In some cases, the unknown may be on the bottom of the ratio.

*Example (3)*  Calculate *x*.

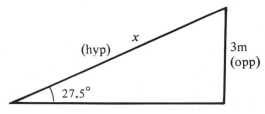

$$\sin 27.5 = \frac{\text{opp}}{\text{hyp}} = \frac{3}{x}$$

Multiply both sides by *x*,

$$x \sin 27.5 = \frac{3}{\cancel{x}} \cancel{x}$$

Divide both sides by sin 27.5,

$$\frac{x \cancel{\sin 27.5}}{\cancel{\sin 27.5}} = \frac{3}{\sin 27.5}$$

$$x = \frac{3}{\sin 27.5} = \frac{3}{0.4617}$$

$$x = 6.498 \text{ m}$$

*Example (4)*

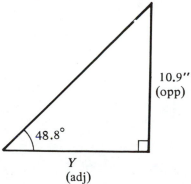

$$\tan 48.8° = \frac{\text{opp}}{\text{adj}} = \frac{10.9}{Y}$$

$$= \frac{10.9}{Y}$$

Multiply both sides by $Y$,

$$Y \tan 48.8° = \frac{10.9}{Y} Y$$

Divide both sides by tan 48.8,

$$Y \frac{\cancel{\tan 48.8}}{\cancel{\tan 48.8}} = \frac{10.9}{\tan 48.8}$$

$$Y = \frac{10.9}{\tan 48.8} = \frac{10.9}{1.142}$$

$$Y = 9.545''$$

## Practice Exercise No. 6

Calculate the lettered sides in the following right triangles.

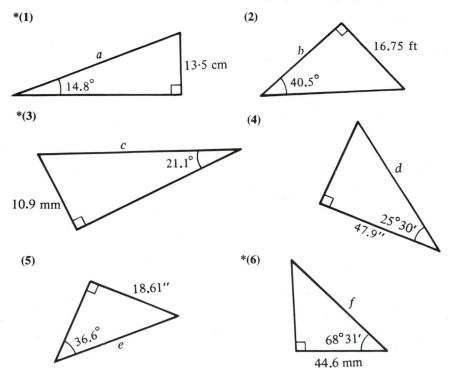

**\*(1)**

$a$

13·5 cm

14.8°

**(2)**

$b$

16.75 ft

40.5°

**\*(3)**

$c$

21.1°

10.9 mm

**(4)**

$d$

47.9″

25°30′

**(5)**

18.61″

36.6°

$e$

**\*(6)**

$f$

68°31′

44.6 mm

## 14.5   PRACTICAL APPLICATIONS
### EXERCISE

Remember that to use sines, cosines, and tangents you must have a right triangle. In some cases, extra lines (----) have been added to figures in order to create right triangles.

**(1)**   The height of the cone in Fig. 14–2 is 12″ and its base angle is 58°. What is the radius of its base?

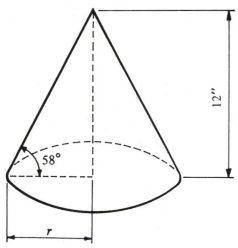

**Fig. 14–2**

**\*(2)**   A pole, 30 m high, is to be secured by 4 wires as shown in Fig. 14–3.

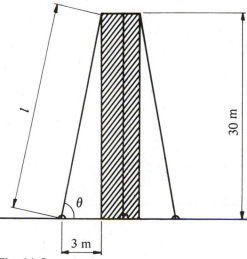

**Fig. 14–3**

      **(a)**   Calculate the length of each wire $l$.

      **(b)**   Find the angle $\theta$ between wire and ground.

**(3)**   The steps of a staircase are 8″ deep and 5.75″ high. Calculate the angle that the bannister rail makes with the horizontal.

**\*(4)**   Figure 14–4 shows the roof of a house. Find the height of the ridge $X$ of the roof above the tops of the walls.

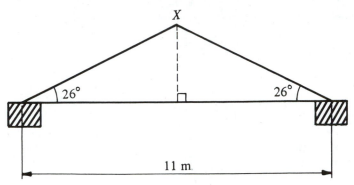

**Fig. 14–4**

**(5)**   A metal plate has three holes drilled as shown in Fig. 14–5. Calculate dimensions $x_1$, $x_2$, $y_1$, and $y_2$.

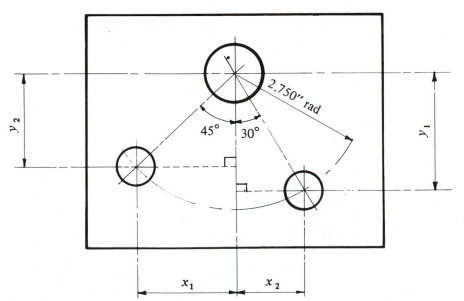

**Fig. 14–5**

**\*(6)** In Fig. 14–6, a circular bar has a flat machined on it (i.e., shaded area is removed).

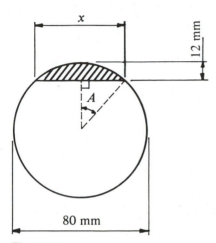

**Fig. 14–6**

(a) Calculate angle $A$, expressed in degrees/minutes.

(b) Find the distance $x$ across the flat face.

**(7)** Figure 14–7 shows details of a groove in a pulley for a vee belt drive. Calculate $x$.

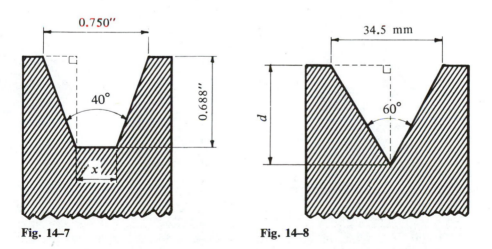

**Fig. 14–7**                          **Fig. 14–8**

**\*(8)** A vee groove is cut in the side of a metal bar as shown in Fig. 14–8. Calculate depth $d$.

(9)  A plate is designed as shown in Fig. 14–9. Calculate the angle α.

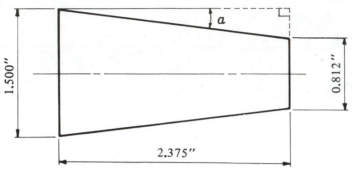

**Fig. 14–9**

*(10)  Figure 14–10 shows a component that is to have two holes drilled in it. From the information given, calculate dimensions A, B, C, and D.

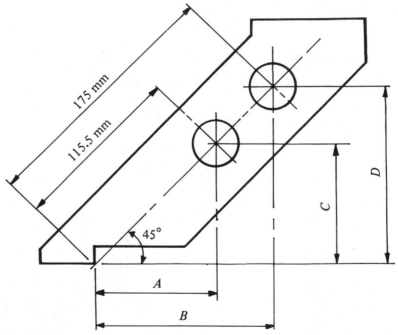

**Fig. 14–10**

(11)  (a)  Calculate the area of the regular hexagon shown in Fig. 14–11.
 (b)  In Fig. 14–12 you are asked to design a wrench to fit the hexagon in Fig. 14–11. Calculate dimension A (allow 0.01″ for clearance on both sides).

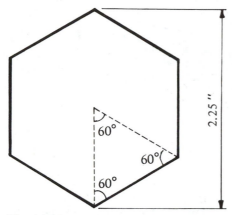

**Fig. 14–11**

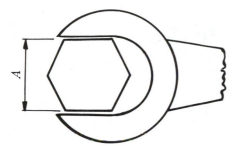

**Fig. 14–12**

**(12)** A disabled person lives in a house that has a doorstep 7" above ground. You are asked to design a ramp that slopes at an angle of 20° to the horizontal. (See Fig. 14–13.) Calculate the length of the ramp $l$ and the distance that it will protrude from the step $d$.

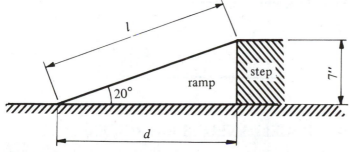

**Fig. 14–13**

**(13)** A taper is to be turned on a bar as shown in Fig. 14–14. Calculate the angle of the taper $\theta$.

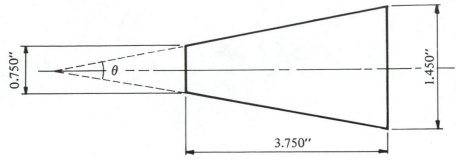

**Fig. 14–14**

**\*(14)** The component shown in Fig. 14–15 is to be produced. Eight holes are to be drilled on a tape-controlled drill press. Five of the holes have centers on an arc of radius 55 mm. From the information given, calculate the dimensions $x_1$, $x_2$, $y_1$, and $y_2$.

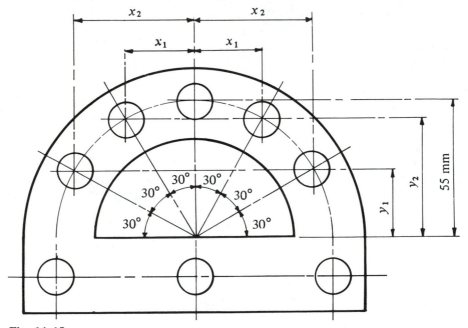

**Fig. 14–15**

## 14.6 MISCELLANEOUS EXERCISES

**(1)** Figure 14–16 shows a two-legged sling being used to lift a mass. The angle between the sling legs is 60° and each leg is 4 yd long.
    **(a)** Find the angle $\alpha$.
    **(b)** Find the distance $x$.

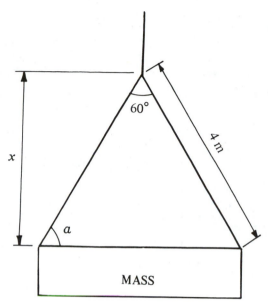

**Fig. 14–16**

*(2)   Determine the dimension *y* shown on the gauge in Fig. 14–17.

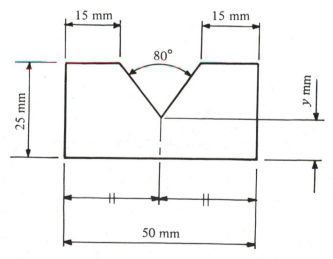

**Fig. 14–17**

*(3)   Calculate the dimension *x* on the gauge shown in Fig. 14–18.

*(4)   From the information given in Fig. 14–19, calculate the angle α.

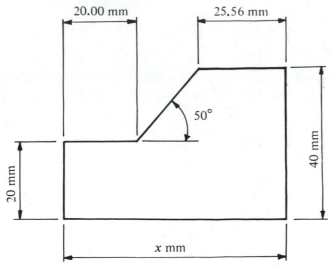

**Fig. 14–18**

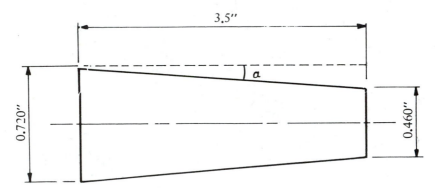

**Fig. 14–19**

**\*(5)**   A folding door is in four equal panels each of width 170 cm. (See Fig. 14–20.)

(a)   What is the maximum width of the door?

(b)   When $AC$ is 300 cm, calculate the perpendicular distance $B$ from $AX$, shown as a dotted line in Fig. 14–20. Then calculate the angle $BAC$.

(c)   Calculate the length $CE$ when angle $CDE = 30°$.

**(6)**   In Fig. 14–21 $ABCD$ is a rectangle. $BE$ and $DF$ are perpendicular to $AC$. $AB = 12''$ and $AD = 7''$.

(a)   Calculate the size of angle $ACD$.

(b)   Find the length of $DF$.

(c)   Find the length of $AE$.

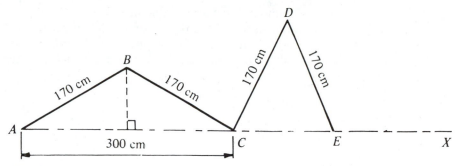

**Fig. 14–20**

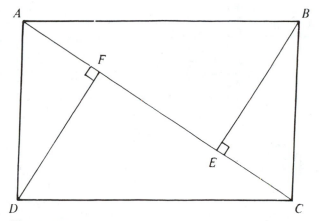

**Fig. 14–21**

(7) Figure 14–22 illustrates a staircase leading from the ground floor to the first floor of a building. The stairs turn through a right angle on the landing *GH*. *BC*, *EG*, and *DF* are vertical posts 1 yd high supporting the stair rails, *AB* and *DE*. Each step of the staircase rises 8″ and has a tread 9″ deep. Both parts of the staircase are 32″ wide.
   (a) What is the height, in feet, above ground of the landing; that is, what is the distance *HY*?
   (b) What is the vertical height, in feet, of *C* above *XY*?
   (c) What angle does the stair rail *DE* make with the horizontal?
   (d) What is the distance, in inches, of *XY*?
   (e) A man who is 5′11″ tall stands upright on the fourth step above the landing *GH*. By how many inches is the top of his head higher than the top of the post *BC*?

(8) Figure 14–23 (not drawn to scale) shows a rectangular box with its lid closed. The box stands with its base horizontal. The dimensions are shown in inches.

(a) (i) Find the volume of the box in cubic inches.
(ii) Find the surface area of the box in square inches.
(iii) Calculate, correct to one decimal place, the length of the longest rod that can be placed in the box and still allow the lid to be closed.

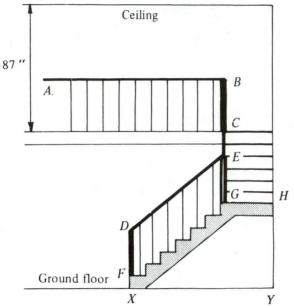

**Fig. 14–22**

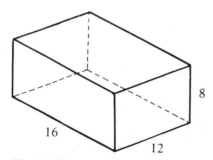

**Fig. 14–23**

(b) Figure 14–24 (not to scale) shows the same box with the lid *ABCD* opened at an angle of 30°. Find the perpendicular height of *B* above *CH*.

(c) If the lid is further opened so that the point *B* is vertically over the midpoint of *CH*, at what angle would the lid now be open?

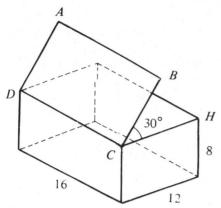

**Fig. 14–24**

*(9)* Figure 14–25 represents a short flight of stairs with a handrail. Each tread is 25 cm and each rise is 18 cm.

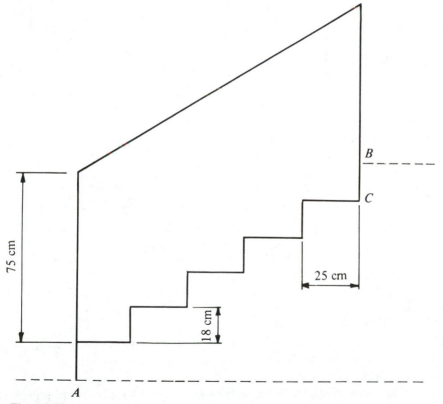

**Fig. 14–25**

(a)   What is the vertical height of *B* above *A*?
(b)   A stair carpet is fitted *A* to *B*. What length of stair carpet is needed?
(c)   The carpet costs $6.20 per meter. What is the cost (to the nearest $0.50) of the carpet required?
(d)   What angle does *AC* make with the horizontal?
(e)   The handrail is parallel to *AC*. What is the angle it makes with the horizontal? Its length?

**(10)**   A cylinder of diameter 2.0″ rests in a vee groove of width 2.4″ as shown in Fig. 14–26. Calculate the height *h*.

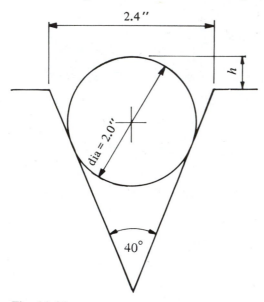

**Fig. 14–26**

**(11)**   If $\sin A = \dfrac{5}{13}$, determine a value for cos *A* and for tan *A* without using trigonometric tables.

**\*(12)**   Refer to Fig. 14–27.
(a)   Determine the dimension *CD*.
(b)   Find the dimension *AC*.
(c)   Calculate the angle α.

**(13)**   Figure 14–28 shows a roller resting in a vee groove. Determine, by calculation, the dimension *H*.

**\*(14)**   A quantity of square blanks of side *L* is to be cut along the line *X–X* as shown in Fig. 14–29 to form a trapezoid and a scrap triangular section.

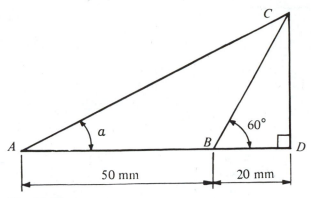

**Fig. 14–27**

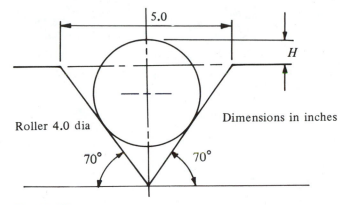

Roller 4.0 dia

Dimensions in inches

**Fig. 14–28**

(a) Calculate the side $L$ of the square.
(b) Find the area of the trapezoid.
(c) Calculate the percentage of scrap.

*(15) Figure 14–30 shows two holes $A$ and $B$ to be marked off for a drill jig. Calculate the dimensions $P$, $Q$, $X$, and $Y$. (*A solution based on measuring the drawing is not acceptable.*)

*(16) The difference in height between the ends of a bar 200 mm long, used to check an angle, is 23.34 mm. What is the angle?

(17) Figure 14–31 shows the part dimensions of a triangular template that is to be checked.
(a) Calculate the vertical height $A$.
(b) Find the dimension $B$.
(c) Calculate the base dimension $C$.

(18) Calculate the angle subtended at the center of a circle of radius 10″ by a chord of length 5.1″.

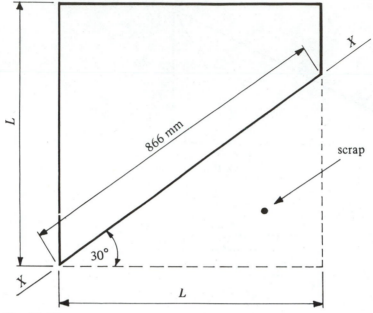

**Fig. 14–29**

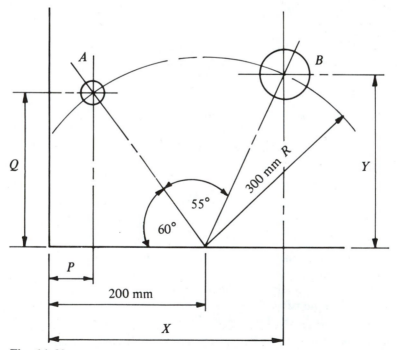

**Fig. 14–30**

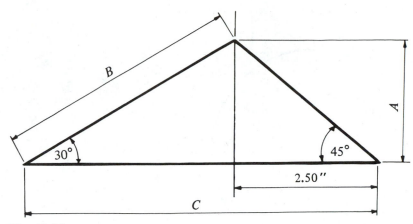

**Fig. 14–31**

**\*(19)** **(a)** Figure 14–32 shows a cylinder of 200 mm in diameter fitting into a 90° symmetrical vee block. Calculate the dimension *B*.

**(b)** A second cylinder is placed in the vee block and the dimension *B* is found to be 40.99 mm. Calculate the diameter of this cylinder.

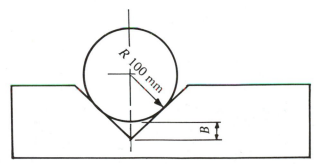

**Fig. 14–32**

**\*(20)** A circular plate has two holes, centers *C* and *D*, marked out on pitch circles of radii 90 mm and 120 mm, respectively, as shown in Fig. 14–33.

**(a)** Calculate the length *CD*.

**(b)** Find the coordinate dimensions *x* and *y*.

**Hint:** Do part (b) first.

**(21)** A template is shown in Fig. 14–34. Calculate the angles θ and ϕ in degrees. Dimensions shown are in inches. The diagram is not to scale. *(A solution based on measuring the drawing is not acceptable.)*

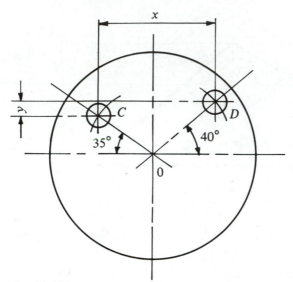

**Fig. 14–33**

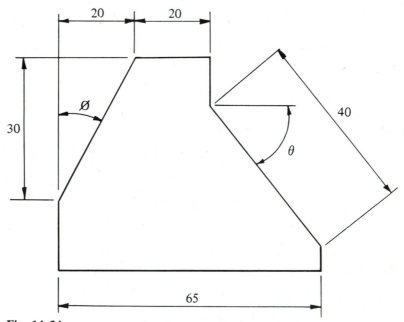

**Fig. 14–34**

# 15

# Trigonometry of Oblique Triangles

**Self-Assessment Test No. 1**

Find:

**(1)** sin 128°

**(2)** cos 147.8°

**(3)** Angle $A$, if sin $A$ = 0.7851

**(4)** Angle $B$, if cos $B$ = 0.7183

**(5)** Angle $C$, if cos $C$ = $-0.6388$

## 15.1 TRIGONOMETRIC VALUES OF OBTUSE ANGLES

### Definitions

An oblique triangle does *not* have a right angle. It may be scalene, isosceles, equilateral, acute-angled, or obtuse-angled. (See Figs. 13–20 to 13–25.)

If the triangle has an obtuse angle, we will need to be able to compute trigonometric values for it, or given its value, we will need to compute the angle. As you will see, we need only the knowledge of how to use trigonometric tables.

### Sine and Cosine of Angles Less Than 180°

Using the trigonometric tables, we previously worked with angles between 0° and 90°. If the angle is between 90° and 180° (obtuse), we will actually look up the **supplement** of the angle.

---

**Calculator Hint**

If you are using a calculator with this chapter, you should refer to Appendix C, Section C.5.

---

## Finding the Value

Remember that supplementary angles are two angles that when added together equal 180°.

***Example (1)***   20° is the supplement of 160°
75° is the supplement of 105°
135° is the supplement of 45°
128°20′ is the supplement of 51°40′

We will look up the supplement of an obtuse angle in Table D2. We will then decide if the value should be positive or negative. For sine, the value is positive ($+$) if the angle is obtuse. For cosine, the value is negative ($-$) if the angle is obtuse. That is,

$$\sin(A) = +\sin(180° - A)$$
$$\cos(B) = -\cos(180° - B)$$

if $A$ and $B$ are between 90° and 180°.

***Example (2)***   $\sin(105°) = +\sin(180° - 105°)$
$= +\sin(75°)$
$= +0.9659$

***Example (3)***   $\cos(128°) = -\cos(180° - 128°)$
$= -\cos(52°)$
$= -0.6157$

***Example (4)***   $\sin(98.3°) = +\sin(180° - 98.3°)$
$= +\sin(81.7°)$
$= +0.9895$

## Practice Exercise No. 1

Find the following.

**(1)**   sin 135°  |  **(2)**   cos 135°
**(3)**   sin 150°  |  **(4)**   cos 159°
**(5)**   sin 28°  |  **(6)**   cos 31°
**(7)**   sin 178°  |  **(8)**   cos 93.7°
**(9)**   sin 151.7°  |  **(10)**   cos 179.7°

## Finding the Angle

For the reverse (what angle yields a particular value), sine and cosine operate slightly differently.

## Cosine

Ignoring the sign, find the angle from Table D2 that corresponds to the value. Then if the sign is negative ($-$), the true angle is its supplement.

***Example (5)*** Find $A$ if $\cos(A) = -0.9848$. The angle that corresponds to 0.9848 under cosine is 10°. Since $-0.9848$ was negative, $A = 180° - 10° = 170°$.

***Example (6)*** Find $B$ if $\cos(B) = 0.3420$. From Table D2, the angle is 70°. Since 0.3420 was positive, $B = 70°$.

## Sine

Again, look up the value in Table D2 to find the angle. However, since the sine is positive for both acute and obtuse angles, it could be the angle from the table *or* its supplement.

***Example (7)*** Find $C$ if $\sin(C) = 0.3256$. From the tables, the angle is 19°. However, $\sin 19° = 0.3256$ and, using the supplement, $\sin 161° = 0.3256$.

***Answer*** 19° or 161° (which is correct can usually be solved by referring to the figure in question).

***Example (8)*** $\sin(D) = 0.9850$

***Answer*** $D = 80.1°$ or 99.9°.

## Practice Exercise No. 2

Find the following angle(s).

**(1)** $\sin(A) = 0.7898$          **(2)** $\cos(B) = 0.8508$

**(3)** $\sin(C) = 0.5736$          **(4)** $\cos(D) = -0.8377$

**(5)** $\sin(E) = 0.8100$          **(6)** $\cos(F) = -0.7500$

**(7)** $\cos(G) = -0.0851$        **(8)** $\cos(H) = 0.7660$

**(9)** $\sin(J) = 0.6814$          **(10)** $\cos(K) = -0.7738$

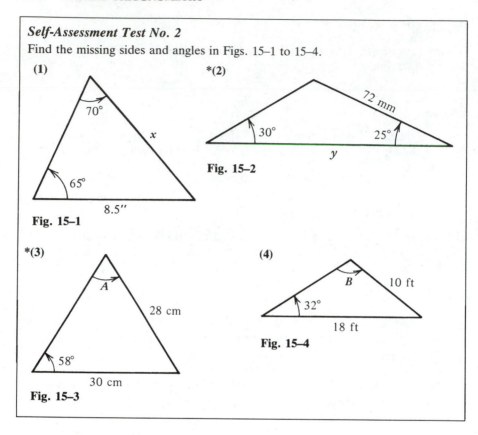

**Self-Assessment Test No. 2**
Find the missing sides and angles in Figs. 15–1 to 15–4.

(1)

Fig. 15–1

*(2)

Fig. 15–2

*(3)

Fig. 15–3

(4)

Fig. 15–4

## 15.2   LAW OF SINES

The formulas we learned in Chapter 14 about opposite side, adjacent side, and hypotenuse no longer apply with oblique triangles. We have two new formulas to replace them.

The simplest to apply is the law of sines.

**Law of Sines:**
For Fig. 15–5,

$$\frac{a}{\sin(A)} = \frac{b}{\sin(B)} = \frac{c}{\sin(C)}$$

In other words, any side divided by the sine of its opposing angle is a constant ratio of that triangle. We will substitute the known values. Then, we will select the pair of ratios that would give us exactly one unknown value.

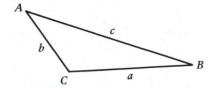

**Fig. 15–5**

*Example (1)*   In Fig. 15–6, find side *a*.

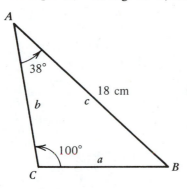

**Fig. 15–6**

Substitute values.

$$\frac{a}{\sin 38°} = \frac{b}{\sin B} = \frac{18}{\sin 100°}$$

Select the first and last ratios.

$$\frac{a}{\sin 38°} = \frac{18}{\sin 100°}$$

$$\frac{a}{0.6157} = \frac{18}{0.9848}$$

Solve for *a*.

$$a = \frac{18(0.6157)}{0.9848}$$

$$a = 11.3 \text{ cm}$$

*Example (2)*   In Fig. 15–6, find *b*. The ratio, $\dfrac{b}{\sin(B)}$, containing *b* has two unknowns, so we must find *B* first.

$$B = 180° - 100° - 38° = 42°$$

Substitute values.

$$\frac{11.3}{\sin 38°} = \frac{b}{\sin 42°} = \frac{18}{\sin 100°}$$

Select the last two ratios.

$$\frac{b}{\sin 42°} = \frac{18}{\sin 100°}$$

Solve for $b$.

$$\frac{b}{0.6691} = \frac{18}{0.9848}$$

$$b = \frac{18(0.6691)}{0.9848}$$

$$b = 12.2 \text{ cm}$$

*Example (3)*   In Fig. 15–7, find angle $A$.

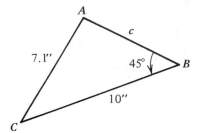

**Fig. 15–7**

$$\frac{10}{\sin(A)} = \frac{7.1}{\sin 45°} = \frac{c}{\sin(C)}$$

$$\frac{10}{\sin(A)} = \frac{7.1}{0.7071}$$

Solve for $\sin(A)$.

$$\sin(A) = \frac{10(0.7071)}{7.1}$$

$$\sin(A) = 0.9959$$

Find $A$ (remember that both the angle and its supplement must be considered).

$$A = 84.8° \quad \text{or} \quad 180° - 84.8°$$

$$\therefore A = 84.8° \quad \text{or} \quad 95.2°$$

*Both* could be correct. If we construct the triangle using the techniques of Chapter 13, we get Fig. 15–8. The arc of radius 7.1 inches from $C$ may cross at $A_1$ or $A_2$. Triangle $CBA_1$ has $A_1 = 84.8°$, and triangle $CBA_2$ has $A_2 = 95.2°$. In an actual application we might be able to discern which was more appropriate based on a sketch or other information.

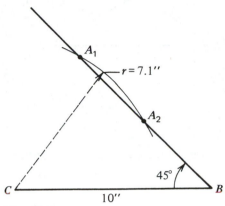

**Fig. 15–8**

## Practice Exercise No. 3

Each problem refers to Fig. 15–9 (not drawn to scale).

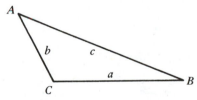

**Fig. 15–9**

(1) $a = 10$, $A = 48°$, $B = 50°$, $b = ?$

(2) $a = 10$, $b = 8.6$, $B = 58°$, $A = ?$

(3) $b = 8.1$, $B = 35.1°$, $C = 120°$, $c = ?$

(4) $a = 12.2$, $c = 16$, $C = 105°$, $A = ?$

(5) $c = 28.7$, $C = 135°$, $B = 25°$, $a = ?$

(6) $b = 8.1$, $c = 9.7$, $B = 38°$, $C = ?$

(7) $b = 9$, $A = 40°$, $C = 98°$, $c = ?$

(8) $c = 158$, $b = 194$, $B = 112°$, $A = ?$

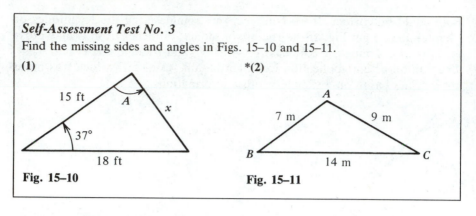

*Self-Assessment Test No. 3*
Find the missing sides and angles in Figs. 15–10 and 15–11.

(1)

*(2)

**Fig. 15–10**

**Fig. 15–11**

## 15.3   LAW OF COSINES

Some oblique triangles cannot be solved using the law of sines. Consider the law of sines for Fig. 15–12.

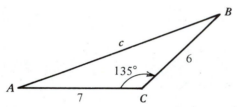

**Fig. 15–12**

$$\frac{6}{\sin(A)} = \frac{7}{\sin(B)} = \frac{c}{\sin 135°}$$

or for Fig. 15–13,

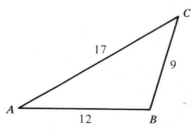

**Fig. 15–13**

$$\frac{9}{\sin(A)} = \frac{17}{\sin(B)} = \frac{12}{\sin(C)}$$

Look back at Section 15.2. You will find that, in every case, we knew at least one side *and* the angle opposite the side. This is a requirement for using the law of sines. For the problem above, we will need the law of cosines.

**Law of Cosines:**

For Fig. 15–14,

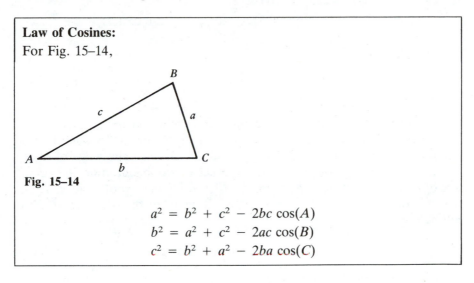

**Fig. 15–14**

$$a^2 = b^2 + c^2 - 2bc \cos(A)$$
$$b^2 = a^2 + c^2 - 2ac \cos(B)$$
$$c^2 = b^2 + a^2 - 2ba \cos(C)$$

Which equation of the three that we use depends on which side or angle we are looking for. For example, the last equation would be used if we were looking for side $c$ or angle $C$ (the letters on the extreme ends of the equation).

***Example (1)*** Find $c$ in Fig. 15–12. Applying the law of cosines, select the appropriate equation.

$$c^2 = a^2 + b^2 - 2ab \cos(C)$$
$$c^2 = 6^2 + 7^2 - 2(6)(7) \cos(135°)$$
$$c^2 = 36 + 49 - 2(6)(7)(-0.7071)$$
$$c^2 = 36 + 49 - 2(-29.70)$$
$$c^2 = 36 + 49 - (-59.4)$$
$$c^2 = 36 + 49 + 59.4$$
$$c^2 = 144.4$$
$$c = 12.02$$

If we now wish to find angle $A$ or $B$, we can easily use the law of sines.

***Example (2)*** Find the angle $B$ in Fig. 15–13. Select the appropriate equation for the problem.

$$b^2 = a^2 + c^2 - 2ac \cos(B)$$
$$17^2 = 9^2 + 12^2 - 2(9)(12) \cos(B)$$
$$289 = 81 + 144 - 216 \cos(B)$$

Solve for $\cos(B)$.

$$289 = 225 - 216 \cos(B)$$

$$289 - 225 = -216 \cos(B) \qquad \text{or} \qquad -216 \cos(B) = 64$$

$$\cos(B) = \frac{64}{-216}$$

$$\cos(B) = -0.2963$$

$$B = 107.3°$$

Notice that we do not have the ambiguity about the angle or its supplement that we had when we used the law of sines. Again, if we wish to know the other angles of Fig. 15–13, we may use the easier-to-apply law of sines.

If you are trying to find all of the angles of an oblique triangle, a good rule of thumb is to find the *largest* angle first, using the law of cosines. Then, use the law of sines for the other angles. Remember that the largest angle will be opposite the largest side.

### Practice Exercise No. 4

For a standard triangle like Fig. 15–14, find the indicated side or angle.

**(1)** $a = 15$, $b = 15$, $C = 38°$, $c = $ ?

**(2)** $a = 10$, $b = 7$, $c = 9$, $A = $ ?

**(3)** $a = 12$, $B = 105°$, $c = 14$, $b = $ ?

**(4)** $a = 12$, $b = 25$, $c = 15$, $B = $ ?

**(5)** $A = 150°$, $b = 53$, $c = 75$, $C = $ ?

**(6)** $a = 142$, $b = 160$, $c = 43$, $C = $ ?

## 15.4 USING TRIGONOMETRY TO FIND THE AREA OF A TRIANGLE

Earlier we used $A = \frac{1}{2}bh$ and Hero's formula to find the area of a triangle.

Now we can add three additional formulas using trigonometry.
The area of Fig. 15–15 can be found with this formula.

$$\text{Area} = \frac{1}{2}ab \sin(C)$$

$$= \frac{1}{2}ac \sin(B)$$

$$= \frac{1}{2}bc \sin(A)$$

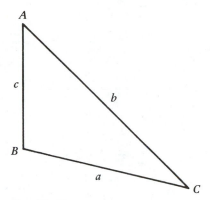

**Fig. 15–15**

***Example*** If $a = 7''$, $b = 8''$, and $C = 39°$ in Fig. 15–15, then

$$\text{Area} = \frac{1}{2}(7)(8) \sin 39°$$

$$= \frac{1}{2}(56)\,(0.6293)$$

$$= 28\,(0.6293)$$

$$= 17.62 \text{ in.}^2$$

### Practice Exercise No. 5

Find the area of the following triangles (refer to Fig. 15–15).

*(1)   $a = 8$ cm, $c = 9$ cm, $B = 25°$

(2)   $c = 12''$, $b = 14''$, $A = 138°$

*(3)   $a = 123$ mm, $b = 281$ mm, $C = 95°$

(4)   $a = 24'$, $b = 75'$, $A = 18°$, $B = 74.9°$

(5)   $a = 18''$, $b = 12''$, $c = 25''$

## 15.5  PRACTICAL APPLICATIONS EXERCISE

(1)   A corner lot is for sale. We know the angle of the corner and the frontage lengths on both sides. Find the length of side $x$, as illustrated in Fig. 15–16.

(2)   Find the area of the lot for sale in Fig. 15–16.

*(3)   Figure 15–17 shows a broadcasting tower, 35 m high, with two supporting guy-wires: $a$ and $b$. How long should the left guy-wire $a$ be?

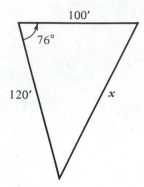

**Fig. 15–16**

*(4)  In Fig. 15–17, how long should the right guy-wire *b* be?

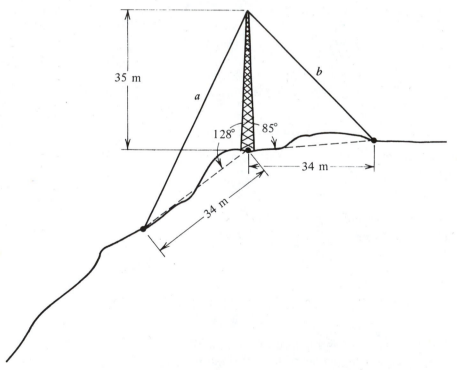

**Fig. 15–17**

*(5)  Find the center-to-center distance between gears *C* and *D* in Fig. 15–18.

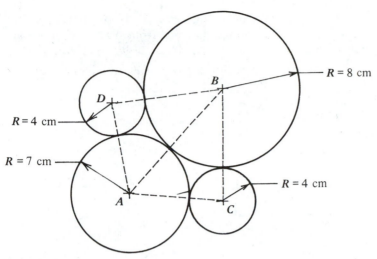

**Fig. 15–18**

**(6)** The ladder leaning against the wall in Fig. 15–19 makes an angle of 60°
with the ground. It touches the wall at the top, creating an angle of 32°
between the ladder and the wall. The wall is *not* vertical or perpendicular with respect to the ground. If the ladder is 30 ft long, how high is
the wall?

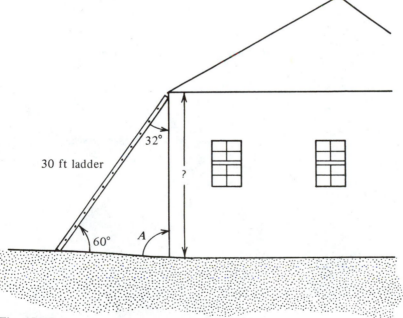

**Fig. 15–19**

*(7)   Figure 15–20 shows a plot of land being surveyed. To check the positions of the stakes, we will measure the diagonal from *A* to *C* when we are through positioning. How long should *AC* be if we are correct?

*(8)   In Fig. 15–20, what is the area of the plot in square meters?

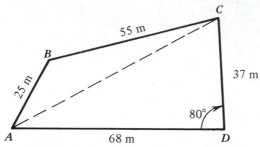

**Fig. 15–20**

# Appendixes

# A

# Scientific Notation

*Self-Assessment Test*

Express the following in scientific notation.

(1)  25.6      (2)  211      (3)  21 000      (4)  197.6

(5)  101 000      (6)  0.752      (7)  0.007 13      (8)  0.000 37

(9)  0.0105      (10)  0.000 001

Convert the following from scientific notation to ordinary form.

(11)  $5.9 \times 10^2$      (12)  $9.7 \times 10^3$      (13)  $4.5 \times 10^{-1}$

(14)  $1.3 \times 10^{-4}$      (15)  $1.1 \times 10^{-2}$

## A.1 NUMBERS IN SCIENTIFIC NOTATION

Any number can be expressed as a value between 1 and 10 multiplied by a power of 10. A number written in this way is said to be in **scientific notation**.

*Example (1)*  $57.8 = 5.78 \times 10^1$

*Example (2)*  $578 = 5.78 \times 100 = 5.78 \times 10^2$

*Example (3)*  $5780 = 5.78 \times 1000 = 5.78 \times 10^3$

*Example (4)*  $0.578 = \dfrac{5.78}{10^1} = 5.78 \times 10^{-1}$      since $\dfrac{1}{10^1} = 10^{-1}$

*Example (5)*  $0.0578 = \dfrac{5.78}{100} = \dfrac{5.78}{10^2} = 5.78 \times 10^{-2}$

As an example of an easy way to remember how to express numbers in scientific notation consider the number 529. To obtain a number between 1 and 10, the decimal point must be moved two places to the left, that is,

**281**

5.2̂9̂. The power of 10 equals the number of places the decimal point is moved to the *left*. Thus,

$$529. = 5.2̂9̂ \times 10^2 \qquad \textbf{\textit{Answer}} \quad 5.29 \times 10^2$$
$$746\ 000 = 7.4̂6000̂ \times 10^5 \qquad \textbf{\textit{Answer}} \quad 7.46 \times 10^5$$
$$6371 = 6.3̂7̂1 \times 10^3 \qquad \textbf{\textit{Answer}} \quad 6.371 \times 10^3$$

When the decimal point has to be moved to the right, the power of 10 is negative. This is used when expressing *numbers less than 1* in scientific notation. In this case, the power of 10 equals the number of places the decimal point is moved to the *right*.

**Example (1)**  $0.519 = 5̂.19 \times 10^{-1}$  **Answer**  $5.19 \times 10^{-1}$
**Example (2)**  $0.000\ 519 = 000̂5̂.19 \times 10^{-4}$  **Answer**  $5.19 \times 10^{-4}$
**Example (3)**  $0.0162 = 01̂.62 \times 10^{-2}$  **Answer**  $1.62 \times 10^{-2}$

## Practice Exercise No. 1

Express the following in scientific notation.

| | | | |
|---|---|---|---|
| **(1)** 777 | **(2)** 8193 | **(3)** 42 | **(4)** 673 000 |
| **(5)** 66.93 | **(6)** 0.000 998 | **(7)** 0.933 | **(8)** 0.0069 |
| **(9)** 8134.9 | **(10)** 89.99 | **(11)** 0.000 007 78 | |
| **(12)** 1333.94 | **(13)** 16.992 | **(14)** 19.004 | **(15)** 8995.4 |
| **(16)** 0.001 | **(17)** 1137 | **(18)** 0.007 71 | **(19)** 0.000 33 |
| **(20)** 1 000 000 | | | |

## Practice Exercise No. 2

Convert the following from scientific notation to ordinary form.

| | | | |
|---|---|---|---|
| **(1)** $5.3 \times 10^2$ | **(2)** $5.9 \times 10^3$ | **(3)** $5.1 \times 10^1$ | **(4)** $6.2 \times 10^3$ |
| **(5)** $1.119 \times 10^5$ | **(6)** $1.79 \times 10^2$ | **(7)** $1.8 \times 10^4$ | **(8)** $1.991 \times 10^1$ |
| **(9)** $2.05 \times 10^{-1}$ | **(10)** $3.14 \times 10^{-3}$ | **(11)** $3.77 \times 10^{-2}$ | **(12)** $1.97 \times 10^{-3}$ |
| **(13)** $1.63 \times 10^{-5}$ | **(14)** $8.96 \times 10^{-1}$ | **(15)** $1.68 \times 10^{-4}$ | **(16)** $1.97 \times 10^{-6}$ |

## A.2 CALCULATIONS WITH SCIENTIFIC NOTATION

A knowledge of the laws of exponents and scientific notation proves to be very useful in certain types of calculations.

*Example (1)* $\dfrac{3000 \times 2000 \times 0.0002}{0.0006} = \dfrac{3 \times 10^3 \times 2 \times 10^3 \times 2 \times 10^{-4}}{6 \times 10^{-4}}$

$$= \dfrac{3 \times 2 \times 2 \times 10^{3+3+(-4)}}{6 \times 10^{-4}}$$

$$= \dfrac{\overset{2}{\cancel{12}} \times 10^2}{\underset{1}{\cancel{6}} \times 10^{-4}}$$

$$= 2 \times 10^{2-(-4)} = 2 \times 10^6$$

*Example (2)* $\dfrac{400 \times 3.5 \times 0.01}{0.02} = \dfrac{\overset{2}{\cancel{4}} \times 10^2 \times 3.5 \times 1 \times \cancel{10}^{-2}}{\underset{1}{\cancel{2}} \times \cancel{10}^{-2}}$

$$= 7.0 \times 10^2$$

## Practice Exercise No. 3

Work out the following expressing the answer in scientific notation.

**(1)** $300 \times 600$

**(2)** $5000 \times 150$

**(3)** $\dfrac{8000}{0.02}$

**(4)** $\dfrac{1600 \times 5000}{40}$

**(5)** $0.01 \times 0.02 \times 0.03$

**(6)** $\dfrac{0.04 \times 600}{0.1}$

**(7)** $\dfrac{50 \times 0.5}{250}$

**(8)** $\dfrac{10\,000 \times 50 \times 100}{0.5 \times 10}$

**(9)** $\dfrac{1.1 \times 60 \times 7.2}{0.2}$

**(10)** $\dfrac{162 \times 35}{0.7 \times 30}$

**(11)** Express 422.6 in scientific notation.

**(12)** Evaluate $\dfrac{0.0027 \times 400}{9000 \times 0.000\,08}$ by first putting each number in scientific notation and then using the laws of exponents. Give your answer in scientific notation.

**(13)** Express the following in decimal form.
  (a) $6.825 \times 10^7$
  (b) $4.351 \times 10^{-4}$

**(14)** Find the value of

$$\dfrac{6 \times 10^3 \times 8 \times 10^{-6}}{32 \times 10^{-3} \times 3 \times 10^6}$$

# B

## Metrics

### B.1 INTRODUCTION

The system of measurements that we usually use (feet, inches, yards, pounds, etc.) is commonly called the English System or the U. S. Customary System. It is based largely on common fractions $\left(\frac{1}{2}, \frac{1}{3}, \frac{1}{4}, \frac{1}{8}, \frac{1}{12}, \frac{1}{16}, \text{etc.}\right)$ and is used officially only in the United States.

The system that the rest of the world uses is called the metric system, or more properly, Système Internationale, or the SI system. It is based on decimal numbers, rarely using fractions. As a result, it is easier to convert from one unit to another, and it is easier to do calculations in general.

The metric system is based on the number 10 and its powers. Measurements of weight, length, capacity, and so on, are formed from a **base** unit, such as the meter, liter, gram, etc., and a prefix to designate the power of ten. For example, the word *centimeter* is formed from the prefix *centi*, meaning $\frac{1}{100}$, and *meter*, which is the base unit for length. Therefore, 1 centimeter is $\frac{1}{100}$ of a meter. That is, 1 meter = 100 centimeters.

The following are the most commonly used of these prefixes.

| Name | Meaning | Abbreviation |
|------|---------|--------------|
| kilo- | 1000 | k |
| hecto- | 100 | h |
| deka- | 10 | da |
| deci- | $\frac{1}{10}$ | d |

centi-                $\dfrac{1}{100}$                 c

milli-                $\dfrac{1}{1000}$                m

In practice, the systems are used separately, and it is rarely necessary to convert from the Customary to the SI form, or from SI to Customary. You will measure, cut, drill, grind, and so forth, from a blueprint written entirely in SI, using SI instruments such as a metric micrometer, metric caliper, meter stick, etc. Actual conversions from one system to another are usually only done by draftsmen using a high degree of precision. Most of us will only need to roughly estimate from one to another because we have developed a "feel" for Customary units. You can develop the same feel for SI units if you remember the following.

· A meter is a little longer than a yard (1.1 yd).

· A liter is a little larger than a quart (1.06 qt).

· A nickel weighs 5 grams.

· A nickel has a diameter of 2 centimeters.

· A millimeter is the thickness of a dime.

· A kilo is about 2.2 pounds.

· A kilometer is about 0.6 miles.

· A metric ton is a little more than a ton.

## B.2   UNITS OF LENGTH: METERS

The meter is the base unit for line measurements such as the length of a house, the height of a wall, or the length of a football field. The kilometer would be used for distances on a highway; the centimeter for dimensions of a textbook or the height of a lamp. Millimeters would be used for very small measurements such as the diameter of a drill-bit or bolt.

The meter is abbreviated by m. The most common units are:

kilometer (km)

meter (m)

centimeter (cm)

millimeter (mm)

In converting from one unit to another, we *multiply* if going to a smaller unit and *divide* if going to a larger unit. The conversion factors you need most frequently are:

$$1 \text{ km} = 1000 \text{ m}$$
$$1 \text{ m} = 100 \text{ cm}$$
$$1 \text{ m} = 1000 \text{ mm}$$
$$1 \text{ cm} = 10 \text{ mm}$$

*Example (1)*   Convert 25.7 m to centimeters.

*Answer*   We are going from meters to centimeters, a smaller unit; therefore, we multiply.

$$25.7 \text{ m} = 25.7 \, (100 \text{ cm})$$
$$= 25 \, 700 \text{ cm}$$

*Example (2)*   Convert 387 mm to centimeters.

*Answer*   We are going to a larger unit.

$$387 \text{ mm} = (387 \div 10) \text{ cm}$$
$$= 38.7 \text{ cm}$$

### Practice Exercise No. 1

Convert the following.

(1)   13 km = ___ m
(2)   8.75 cm = ___ mm
(3)   285 mm = ___ m
(4)   31 m = ___ cm
(5)   27 051 m = ___ km
(6)   2.7 m = ___ mm
(7)   0.015 mm = ___ cm
(8)   28 cm = ___ m

## B.3   UNITS OF AREA: SQUARE METERS

$$1 \text{ m}^2 = 10 \, 000 \text{ cm}^2$$
$$= 1 \, 000 \, 000 \text{ mm}^2$$
$$1 \text{ km}^2 = 1 \, 000 \, 000 \text{ m}^2$$
$$1 \text{ cm}^2 = 100 \text{ mm}^2$$

*Example (1)*   Convert 56.3 mm² to square centimeters.

*Answer*   We are going from square millimeters to square centimeters, a larger unit; therefore, we divide.

$$56.3 \text{ mm}^2 = (56.3 \div 100) \text{ cm}^2$$
$$= 0.563 \text{ cm}^2$$

*Example (2)*   Convert 128.7 km² to square meters.

*Answer*   We are going from square kilometers to square meters, a smaller unit; therefore, we multiply.

$$128.7 \text{ km}^2 = 128.7 \ (1\ 000\ 000) \text{ m}^2$$
$$= 128\ 700\ 000 \text{ m}^2$$

### Practice Exercise No. 2

Convert the following.

**(1)**   28 m² = ___ cm²

**(2)**   187 581 m² = ___ km²

**(3)**   87 151 cm² = ___ mm²

**(4)**   87.5 mm² = ___ cm²

**(5)**   28.7 km² = ___ m²

**(6)**   1 357 246 mm² = ___ m²

## B.4   UNITS OF VOLUME: CUBIC METERS

$$1 \text{ m}^3 = 1\ 000\ 000 \text{ cm}^3 \text{ (cc)}^*$$
$$= 1\ 000\ 000\ 000 \text{ mm}^3$$
$$1 \text{ km}^3 = 1\ 000\ 000\ 000 \text{ m}^3$$
$$1 \text{ cm}^3 = 1000 \text{ mm}^3$$
$$1 \text{ dm}^3 = 1000 \text{ cm}^3$$

*Example (1)*   Convert 2075 mm³ to cubic centimeters.

*Answer*   We are going from a smaller unit to a larger unit; therefore, we divide.

$$2075 \text{ mm}^3 = (2075 \div 1000) \text{ cm}^3$$
$$= 2.075 \text{ cm}^3$$

*Example (2)*   Convert 5.1 cc to cubic millimeters.

*Answer*   We are going to a smaller unit; therefore, we multiply.

$$5.1 \text{ cc} = 5.1 \ (1000) \text{ mm}^3$$
$$= 5100 \text{ mm}^3$$

### Practice Exercise No. 3

Convert the following.

**(1)**   25 781 cm³ = ___ m³

**(2)**   0.581 km³ = ___ m³

**(3)**   12 cc = ___ mm³

**(4)**   1075 mm³ = ___ cm³

**(5)**   2.7 m³ = ___ cm³

**(6)**   1 cm³ = ___ dm³

---

* The abbreviation cm³ is popularly abbreviated as cc in the United States, particularly in medical fields.

## B.5 UNITS OF LIQUID CAPACITY: LITERS

The liter (l or L) measures liquid capacity. Liquids such as milk, wine, or gasoline are sold retail by the liter. Large containers, such as the drums of liquid chemicals used by industries, are designated as 2 hectoliters. A hecto-liter (hl) is 100 liters. Very small units, such as those used with medications, are measured in milliliters (ml). A liter is 1000 ml.

$$1 \text{ liter} = 1000 \text{ milliliters (ml)}$$
$$100 \text{ liters} = 1 \text{ hectoliter (hl)}$$

*Example (1)*   Convert 7.8 L to milliliters.

*Answer*   We are going to a smaller unit; therefore, we multiply.

$$7.8 \text{ L} = 7.8 \ (1000) \text{ ml}$$
$$= 7800 \text{ ml}$$

### Comparison of Volume and Capacity

If a rectangular container measured 1 decimeter on each side (a cubic deci-meter), it would contain *exactly* 1 liter. This allows us to convert easily from volume measurements to liquid measurement.

$$1 \text{ dm}^3 = 1 \text{ L}$$
$$= 1000 \text{ cm}^3$$
$$1 \text{ m}^3 = 1000 \text{ L} = 1 \text{ kl}$$
$$1 \text{ cm}^3 = 1 \text{ ml}$$

(You might want to compare the equivalent conversions from volume to capacity in the Customary System. Did you know that 1 gal = 231 in.$^3$ or that 1 ft$^3$ = 7.48 gal? And those are only approximations; the metric con-versions are exact.)

*Example (2)*   Convert 275 cm$^3$ to milliliters.

*Answer*   Since 1 cm$^3$ = 1 cc = 1 ml, 275 cm$^3$ = 275 ml.

*Example (3)*   Convert 2.7 m$^3$ to liters.

*Answer*   Since we are going from a larger unit to a smaller unit, we multiply.

$$2.7 \text{ m}^3 = 2.7 \ (1000) \text{ L}$$
$$= 2700 \text{ L}$$

### Practice Exercise No. 4

Convert the following.

(1)   287 ml = ___ L                    (2)   28 L = ___ ml

(3)   5 kl =  __ L                                      (4)   287 ml =  __ cm³

(5)   28 L =  __ dm³                                    (6)   7851 ml =  __ L

## B.6   UNITS OF WEIGHT: GRAMS

The gram (g) is used to measure weight or mass. It is used to weigh small objects such as packaged cereal, canned vegetables, or in mailing packages. Very small quantities such as medication would be measured in milligrams. Bulk items such as potatos, iron bars, meat, lawn fertilizer, or cattle would be weighed in kilograms.

$$1 \text{ gram (g)} = 1000 \text{ milligrams (mg)}$$
$$1000 \text{ grams} = 1 \text{ kilogram (kg)}$$
$$1000 \text{ kilogram} = 1 \text{ metric ton}$$

---

**Note:**   *The kilogram is popularly called a **kilo** (plural is **kilos**).*

---

***Example (1)***   Convert 57 g to milligrams.

***Answer***   57 g = 57 (1000) mg
             = 57 000 mg

***Example (2)***   Convert 3875 g to kilograms.

***Answer***   3875 g = (3875 ÷ 1000) kg
               = 3.875 kg

### Practice Exercise No. 5

Convert the following.

(1)   1250 mg =  __ g                                   (2)   225 g =  __ kg

(3)   2785 kg =  __ metric tons

(4)   25 kg =  __ g                                     (5)   2.7 g =  __ mg

(6)   500 mg =  __ g                                    (7)   2.2 kilo =  __ g

# C
# The Calculator

This appendix is referred to in the text at points where it is helpful to use calculators.

There are a variety of hand-held calculators on the market, offering a variety of functions. Since this book uses squares, square roots, and trigonometric functions, we will restrict our discussion to "scientific calculators," that is, to those with $\boxed{x^2}$, $\boxed{\sqrt{x}}$, $\boxed{\sin}$, $\boxed{\cos}$, $\boxed{\tan}$, among other keys. These calculators can be classified according to the *order* in which operations must be entered. The two basic types are RPN calculators and algebraic-notation calculators (ALG or AEN are common abbreviations for this type).

RPN calculators, such as Hewlett-Packards, some Sinclairs, and a few others, can generally be recognized by the presence of an $\boxed{\text{enter} \uparrow}$ or $\boxed{\uparrow}$ key. RPN calculators will not be covered in this appendix since their operation is more complex than the algebraic-notation calculators; however, the manuals provided with these calculators are, of necessity, very complete.

Algebraic-notation calculators, such as the TI-30 by Texas Instruments, are designed so that you may enter numbers and operations in approximately the same order that you would write them. Some of these calculators, **Type A**, preserve the established order of operations (multiplication and division are done before addition and subtraction). Some others, **Type B**, do each operation immediately after it is entered. The difference depends on how many **registers** the calculator has. Type A calculators must have at least three registers. A quick test is to enter the following key strokes.

$$\boxed{2}\boxed{\times}\boxed{3}\boxed{+}\boxed{4}\boxed{\times}\boxed{5}\boxed{=}$$

If your result is 26, then you have a Type A.

If your result is 50, then you have a Type B.

The Type A calculator **reserved** the addition until all consecutive multiplications and divisions had been done (i.e., until another addition or = was keyed).

We will assume the use of Type A, algebraic-notation, scientific calculators. If you are using a Type B, we recommend that no more than *one* operation at a time be done on the calculator, inserting each result in your problem as you get it.

## C.1 BASIC OPERATIONS

### Order of Operations

As already mentioned, some operations are reserved; others are immediately performed. This is the result of **registers** (operating memories). The **X-register** is the displayed result, or the last number entered. The **Y-register** is the next-to-the-last number entered. The other registers are used to "reserve" numbers and operations until they are called up.

The best way to remember how each operation works is to divide the operations into levels.

### Level 1

$\boxed{+}$, $\boxed{-}$, and $\boxed{=}$ are all level 1 operations. When any one of them is keyed, *all* previous operations are performed and the result is stored in the X-register. In the case of $\boxed{+}$ and $\boxed{-}$, the result will be immediately shifted to the Y-register if we start entering the second number (addend or subtrahend). In the case of $\boxed{=}$, it will be lost unless we enter an operation after it (such as $\boxed{+}$, $\boxed{-}$, etc.).

### Level 2

$\boxed{\times}$, $\boxed{\div}$, $\boxed{y^x}$, and $\boxed{\sqrt[x]{y}}$ are level 2 operations. When one of them is keyed, the operation is reserved until the next number is entered and the next operation is keyed. At that point, the level 2 operation is performed on the X- and Y-registers, that is, the numbers immediately preceding and following the level 2 operation.

### Level 3

$\boxed{\sin}$, $\boxed{\cos}$, $\boxed{\tan}$, their inverses, $\boxed{e^x}$, $\boxed{\log}$, $\boxed{x^2}$, $\boxed{\sqrt{x}}$, $\boxed{\dfrac{1}{x}}$, and $\boxed{x!}$ are level 3 operations. They are performed immediately on the number in the X-register (display) only.

To see how these three levels work, consider the following examples.

*Example (1)* 2(3) + 7(8)(5)

| Keys | Display Results | Comments |
|---|---|---|
| $\boxed{2}$ | 2 | |
| $\boxed{\times}$ | 2. | Decimal appears as result of operation. |
| $\boxed{3}$ | 3 | Level 2 operation is reserved. |
| $\boxed{+}$ | 6. | All outstanding operations are performed. |
| $\boxed{7}$ | 7 | |
| $\boxed{\times}$ | 7. | |
| $\boxed{8}$ | 8 | Level 2 operation is reserved. |
| $\boxed{\times}$ | 56. | Level 2 operation only is performed, not the $\boxed{+}$. |
| $\boxed{5}$ | 5 | Level 2 operation is reserved. |
| $\boxed{=}$ | 286. | All outstanding operations are performed. |

*Example (2)* $2(8^2)(3)$

| Keys | Display Results | Comments |
|---|---|---|
| $\boxed{2}$ | 2 | |
| $\boxed{x}$ | 2. | Decimal appears. |
| $\boxed{8}$ | 8 | Level 2 operation is reserved. |
| $\boxed{x^2}$ | 64. | Only the displayed number is squared. |
| $\boxed{x}$ | 128. | Previous $\boxed{x}$ is performed. |
| $\boxed{3}$ | 3 | Level 2 (second $\boxed{x}$) is reserved. |
| $\boxed{=}$ | 384 | All operations are performed. |

## Practice Exercise No. 1

Using a calculator, do the following calculations.

**(1)** (23.8)(31.6) + (105.6)(4.09)

**(2)** (286 000)(0.311) − (0.886)(1962)

**(3)** (60.7)(1059)(237)

**(4)** (988)(0.0556)(66.1) + 1.085

**(5)** (15.06)($\pi$)(625)(2.93)(17.22)

**(6)**   $(0.1153) \div (70.3) + (8.5)(7.1)$

**(7)**   $(0.774^2)(\sqrt{11.47})$

**(8)**   $(54.23^2)(81.7)$

**(9)**   $(7.90)(7.02)(54.8^2)(3)$

**(10)**   $\sqrt{35.6}\ \sqrt{71.8}$

## Division

Divisions, as we usually write them, require special handling. Consider Example (3) below. The long line indicates division by 5, but it also means that the entire top must be calculated first. We do **not** allow the calculator to **reserve** operations. We will force the top calculation with the $\boxed{=}$ key.

*Example (3)*   $\dfrac{2(80) + 90}{5}$

| Keys | Display Results | Comments |
|------|-----------------|----------|
| $\boxed{2}$ | 2 | |
| $\boxed{\times}$ | 2. | |
| $\boxed{80}$ | 80 | $\boxed{\times}$ is reserved. |
| $\boxed{+}$ | 160 | All operations are performed. |
| $\boxed{90}$ | 90 | |
| $\boxed{=}$ | 250. | Forces the top to be totally calculated. |
| $\boxed{\div}$ | 250. | |
| $\boxed{5}$ | 5 | $\boxed{\div}$ is reserved. |
| $\boxed{=}$ | 50. | All operations are performed. |

An alternate way to do Example (3) is to rewrite the problem in linear form. We use parentheses to express the order of operations intended, instead of top and bottom positions. Hence

$$\frac{2(80) + 90}{5} = (2(80) + 90) \div 5$$

We then regard the closing of the parentheses as the $\boxed{=}$ key. Or, if your calculator has a parentheses key, use it to key in the problem.

*Example (4)*   $\dfrac{2(8) + 7}{7(5)}$

We can rewrite this as

$$(2(8) + 7) \div (7(5))$$

Unless you have parentheses on your calculator, the calculations *after the division sign* should be performed first and stored (in memory or on paper) before the first part of the equation can be performed.

> **Hint**
>
> It is better in the long run *not* to chain operations like this on your calculator without copying down the results. Calculate $2(8) + 7$ and write it down. Calculate $7(5)$ and write it down under the last result:
>
> $$\frac{23}{35}$$
>
> Then divide 23 by 35.

*Answer*   0.657 142 9

## Practice Exercise No. 2

Using your calculator, do the following calculations.

(1) $\dfrac{25.6(2.4)}{5.6}$

(2) $\dfrac{38.7 + 3.2(8.7)}{7.35}$

(3) $\dfrac{3.7(4.8) - 2.7(0.185)}{6.2 + 3.9}$

(4) $\dfrac{1.25 + 1.12}{97.5(2.3)}$

(5) $\dfrac{28.32 + 9.43}{2.03 - 1.84}$

(6) $\dfrac{226.7 - 9.43}{210.6 - 14.96}$

(7) $\dfrac{26.6}{3.9} - \dfrac{23.0}{6.94}$

(8) $\dfrac{5(96 - 32)}{9}$

(9) $r = \dfrac{r_1 r_2}{r_1 + r_2}$, where $r_1 = 69.4$ and $r_2 = 73.1$

**(10)**  $F = \dfrac{9C}{5} + 32$, where $C = 38$

## Scientific Notation

Most scientific calculators will shift to scientific notation (see Appendix A) if the result of the calculation is too small or too large to display. Some, LCD calculators in particular, will shift to scientific notation if the result is less than 1, or too large.

Do the following two problems on your calculator for a demonstration of how your calculator will display scientific notation.

**(1)**  99 999 999 $\boxed{x^2}$ results in $\boxed{9.999\ 998\quad 15}$, which means $9.999\ 998 \times 10^{15}$ or 9 999 998 000 000 000.

**(2)**  0.000 000 66 $\boxed{x^2}$ results in $\boxed{4.356\quad -13}$, which means $4.356 \times 10^{-13}$ or 0.000 000 000 000 435 6.

## Rounding

Since the calculator gives you 8 or 10 digits as a result of calculations, you will need to round your answers off to 3 or 4 digits before comparing them to the answer key or using the results. Few instruments or machines are built to use 10 digits of accuracy.

**The following sections are about special types of problems that you may solve on your calculator. The text will refer you back to them as you need them. You may also cover them now in preparation for that need.**

## C.2  FRACTIONS WITH A CALCULATOR

### Changing Fractions to Decimals

Divide the top by the bottom.

$$\dfrac{3}{4}: \qquad 3 \div 4 \ = 0.75$$

$$\dfrac{17}{16}: \qquad 17 \div 16 = 1.0625$$

In some cases the decimal value is only an approximation. This is usually the case if the result uses the full display.

$$\dfrac{2}{3}: \qquad 2 \div 3 = 0.666\ 666\ 666\ 7$$

$$\frac{7}{13}: \qquad 7 \div 13 = 0.538\ 461\ 538\ 5$$

On some calculators, this may mean that $\frac{7}{13} \times (13\ 000)$ does not result in 7 000, but some approximate value such as 6 999.999 998.

## Mixed Numbers

Mixed numbers can be changed on ALG calculators by remembering that

$$2\frac{3}{8} = 2 + 3 \div 8 = 2.375$$

## Changing Decimals to Fractions

### An Exact Answer

$$0.2508 = \frac{2508}{10\ 000}$$

then reduce to lowest terms (dividing by 4); thus,

$$0.2508 = \frac{627}{2500}$$

### An Approximate Answer

We can use the calculator to convert decimals to any desired denominator. Quite often we need an answer to the nearest 32nd or 64th, but (as above) it does not reduce to 32nds or 64ths. Therefore, multiply the decimal by the desired denominator, round to the nearest whole number, and use this answer as the numerator.

*Example (1)*   $0.2508 = \frac{?}{32}$

$$(0.2508)(32) = 8.0256 \rightarrow 8 \text{ (rounded)}$$

$$\therefore 0.2508 = \frac{8}{32} \qquad \text{(approximately)}$$

*Example (2)*   $5.7899 = 5\frac{?}{64}$

$$(0.7899)(64) = 50.5536 \rightarrow 51 \text{ (rounded)}$$

$$\therefore 5.7899 = 5\frac{51}{64}$$

## Practice Exercise No. 3

Convert to 4-place decimals.

(1) $\dfrac{3}{8}$

(2) $\dfrac{7}{5}$

(3) $\dfrac{3}{7}$

(4) $\dfrac{5}{9}$

(5) $2\dfrac{7}{8}$

(6) $15\dfrac{1}{3}$

(7) $33\dfrac{1}{3}$

(8) $12\dfrac{1}{2}$

Convert the following decimals to fractions.

(9) $0.75 = \dfrac{?}{4}$

(10) $0.751 = \dfrac{?}{16}$

(11) $0.8518 = \dfrac{?}{16}$

(12) $0.915 = \dfrac{?}{32}$

(13) $2.791 = 2\dfrac{?}{5}$

(14) $16.789\ 34 = 16\dfrac{?}{64}$

(15) $15.759 = 15\dfrac{?}{64}$

(16) $28.71 = 28\dfrac{?}{16}$

## C.3   SQUARES AND SQUARE ROOTS

Remember that $\boxed{x^2}$ and $\boxed{\sqrt{x}}$ were level-3 operations and are performed on whatever number appears in the display. (Also remember that $\boxed{\sqrt{x}}$ does not work on negative numbers.) It is a good idea to keep a close watch on the display when using them. The $\boxed{\sqrt{x}}$ function can be used to replace the square-root table.

### Pythagorean Theorem

One use you will have for the square and square-root keys is in the application of the Pythagorean theorem.

*Example (1)*

$$c = \sqrt{(7.81)^2 + (8.5)^2}$$

Since these are level-3 operations, the keying is direct, except that the $\boxed{=}$ key is used to complete the squares and the sum before the square root is used.

$$7.81\boxed{x^2} + 8.5\boxed{x^2}\boxed{=}\boxed{\sqrt{x}}$$

*Answer*   11.543

*Example (2)*

$$a = \sqrt{(12.1)^2 - (8.3)^2}$$

Again, with a subtraction, the operation is direct, except for the $\boxed{=}$ before $\boxed{\sqrt{x}}$.

*Answer*   8.805

If there is a calculation to be done before the square *or* the square root, the equal key must generally be used.

*Example (3)*   Find $(8.83 - 3.1)^2$. The order should be

$$(8.83 - 3.1)^2 = (5.73)^2 = 32.8329$$

The keystrokes are

$$8.83 \boxed{-} 3.1 \boxed{=} \boxed{x^2}$$

### Practice Exercise No. 4

Calculate the following using your calculator.

(1)   $a = \sqrt{14^2 - 7.5^2}$          (2)   $c = \sqrt{7.3^2 + 4.9^2}$

(3)   $b = \sqrt{21.3^2 - 16.1^2}$         (4)   $x = \sqrt{19.5^2 - 14^2}$

(5)   $(8.75 - 3.12)^2$               (6)   $\sqrt{(2.1)(7.3)(8.4)^2}$

(7)   $(\sqrt{81}\,\sqrt{7} - 3)^2$            (8)   $\sqrt{7^2 + (10 + 8)^2}$
    (**Hint:** Do $\sqrt{7} - 3$ first.)

## C.4   USING TRIGONOMETRIC FUNCTIONS: ANGLES LESS THAN 90°

Most calculators will do trigonometric functions for degrees, radians, and sometimes grads. We will always work in degrees. Some calculators use a "D-R-G" switch to set the state. Others assume **degrees** are being used unless you change the state with a special key. A few assume **radians** unless you change the state. Read your manual and make sure that you have your calculator set for degrees.

Remember that $\boxed{\sin}$, $\boxed{\cos}$, and $\boxed{\tan}$ are level-3 operations. They perform the calculation immediately on the displayed number. Notice also that they are entered on the calculator in the reverse order from the order in which they are read.

*Example (1)*   Find cos 45°. We key in

$$45 \boxed{\cos}$$

*Answer*   0.707 106 781 2

*Example (2)*   Find 28.7 sin 38°. We key in

$$28.7 \; \boxed{\times} \; 38 \; \boxed{\text{sin}} \; \boxed{=}$$

*Answer*   17.669 484 34

(Sin, cos, and tan are not as fast as other operations. Wait until $\boxed{\text{sin}}$ is finished before keying $\boxed{=}$.) To go from value to angle, we use either $\boxed{\text{arc}}$, $\boxed{\text{inv}}$, or $\boxed{\text{2nd}}$. The $\boxed{\text{2nd}}$ is occasionally not labeled, but indicated as an unmarked, colored key (gold, blue, or white). *One* of these keys must *precede* the trigonometric key. If you have an $\boxed{\text{arc}}$ or $\boxed{\text{inv}}$, it *must* be used, even if you also have a $\boxed{\text{2nd}}$ key. Calculators that require the use of $\boxed{\text{2nd}}$ will have $\cos^{-1}$, $\sin^{-1}$, and $\tan^{-1}$ printed above $\boxed{\text{cos}}$, $\boxed{\text{sin}}$, and $\boxed{\text{tan}}$ and will not have $\boxed{\text{inv}}$ at all.

*Example (3)*   If cos(A) = 0.699 66, find *A*. The procedure is

$$0.699 \; 66 \; \boxed{\text{inv}} \; \boxed{\text{cos}}$$

*Answer*   45.6° (to the nearest 0.1°)

*Example (4)*   If tan(A) = 2.759, find *A*. Use the keystrokes

$$2.759 \; \boxed{\text{inv}} \; \boxed{\text{tan}}$$

*Answer*   70.08°

### Practice Exercise No. 5

Find the missing value or angle.

(1)   cos 28.7° = ___          (2)   sin 88.7° = ___

(3)   tan 23° = ___          (4)   tan 76.8° = ___

(5)   cos(A) = 0.917 06          (6)   sin(B) = 0.218 14

(7)   tan(C) = 0.2135          (8)   tan(D) = 4.5687

## C.5   USING TRIGONOMETRIC FUNCTIONS: ANGLES BETWEEN 90° AND 180°

We will restrict our coverage to angles less than 180° (since triangles, rectangles, etc., have angles less than 180°). We will also restrict our coverage to sine and cosine (which are used in Chapter 15).

Almost all calculators will find the value of trigonometric expressions like cos 128° or sin 147°, if you key in the angle and the appropriate function.

However, note the results of the following pairs of values when you key them into your calculator:

**(1)** $\cos 128°$, $\cos 52° \rightarrow -0.6157$, $+0.6157$

**(2)** $\sin 147°$, $\sin 33° \rightarrow +0.5446$, $+0.5446$

The values in (2) are the same, and those in (1) are additive inverses, differing only in sign.

From these and similar examples we could observe that

$$\cos(A) = -\cos(180° - A)$$
$$\sin(A) = \sin(180° - A)$$

This will help us in doing inverse trigonometric functions.

**Example (1)**   Find $A$ if $\cos(A) = -0.8829$. If $\cos(A)$ is negative, then $A$ must be between 90° and 180°. If you key $-0.8829$ into the calculator (generally $0.8829$ $\boxed{+/-}$ ) and then key $\boxed{\text{inv}}$ $\boxed{\cos}$ , you will get a result of 152°.

**Answer**   $A = 152°$

**Example (2)**   Find $B$ if $\cos(B) = 0.7771$. If $\cos(B)$ is positive, then $B$ must be between 0° and 90°. If you key $0.7771$ and $\boxed{\text{inv}}$ $\boxed{\cos}$ , you will get a result of 39°.

**Answer**   $B = 39°$

**Example (3)**   Find $C$ if $\sin(C) = 0.9781$. If $\sin(C)$ is positive, then $C$ might be either less than 90° or more than 90°. If you key $0.9781$ and $\boxed{\text{inv}}$ $\boxed{\sin}$ , you will get a result of 78°. However, we should also consider $180° - 78° = 102°$.

$$\sin(78°) = 0.9781 \quad \text{and} \quad \sin(102°) = 0.9781$$

**Answer**   $C = 78°$  or  $102°$

## Summary

·   Your calculator will give you the correct *value* for either $\boxed{\cos}$ or $\boxed{\sin}$ .

·   Your calculator will give you the correct angle in the range of 0° to 180° for $\boxed{\text{inv}}$ $\boxed{\cos}$ .

·   Two angles are possible for $\boxed{\text{inv}}$ $\boxed{\sin}$ in the range of 0° to 180°. The calculator will give you only the one less than 90°. The other possibility is 180° minus that angle. There is no way to identify which is correct without observing the figure that the angle comes from.

## Practice Exercise No. 6

Find the missing angles (to the nearest 0.1°).

(1) $\sin(A) = 0.5781$

(2) $\cos(A) = -0.7841$

(3) $\cos(A) = -0.9987$

(4) $\sin(A) = 0.1281$

(5) $\cos(A) = 0.4871$

(6) $\cos(A) = 0.1121$

(7) $\sin(A) = 0.9871$

(8) $\cos(A) = -0.7891$

(9) $\sin(A) = 0.0181$

(10) $\sin(A) = 0.7891$

# D

**TABLE D1   Square Roots (1.000 − 9.999)**

| N | 0 | 1 | 2 | 3 | 4 | 5 | 6 | 7 | 8 | 9 | 1 | 2 | 3 | 4 | 5 | 6 | 7 | 8 | 9 |
|---|---|---|---|---|---|---|---|---|---|---|---|---|---|---|---|---|---|---|---|
| 1.0 | 1.000 | 1.005 | 1.010 | 1.015 | 1.020 | 1.025 | 1.030 | 1.034 | 1.039 | 1.044 | 0 | 1 | 1 | 2 | 2 | 3 | 3 | 4 | 4 |
| 1.1 | 1.049 | 1.054 | 1.058 | 1.063 | 1.068 | 1.072 | 1.077 | 1.082 | 1.086 | 1.091 | 0 | 1 | 1 | 2 | 2 | 3 | 3 | 4 | 4 |
| 1.2 | 1.095 | 1.100 | 1.105 | 1.109 | 1.114 | 1.118 | 1.122 | 1.127 | 1.131 | 1.136 | 0 | 1 | 1 | 2 | 2 | 3 | 3 | 4 | 4 |
| 1.3 | 1.140 | 1.145 | 1.149 | 1.153 | 1.158 | 1.162 | 1.166 | 1.170 | 1.175 | 1.179 | 0 | 1 | 1 | 2 | 2 | 3 | 3 | 3 | 4 |
| 1.4 | 1.183 | 1.187 | 1.192 | 1.196 | 1.200 | 1.204 | 1.208 | 1.212 | 1.217 | 1.221 | 0 | 1 | 1 | 2 | 2 | 2 | 3 | 3 | 4 |
| 1.5 | 1.225 | 1.229 | 1.233 | 1.237 | 1.241 | 1.245 | 1.249 | 1.253 | 1.257 | 1.261 | 0 | 1 | 1 | 2 | 2 | 2 | 3 | 3 | 4 |
| 1.6 | 1.265 | 1.269 | 1.273 | 1.277 | 1.281 | 1.285 | 1.288 | 1.292 | 1.296 | 1.300 | 0 | 1 | 1 | 2 | 2 | 2 | 3 | 3 | 4 |
| 1.7 | 1.304 | 1.308 | 1.311 | 1.315 | 1.319 | 1.323 | 1.327 | 1.330 | 1.334 | 1.338 | 0 | 1 | 1 | 2 | 2 | 2 | 3 | 3 | 3 |
| 1.8 | 1.342 | 1.345 | 1.349 | 1.353 | 1.356 | 1.360 | 1.364 | 1.367 | 1.371 | 1.375 | 0 | 1 | 1 | 1 | 2 | 2 | 3 | 3 | 3 |
| 1.9 | 1.378 | 1.382 | 1.386 | 1.389 | 1.393 | 1.396 | 1.400 | 1.404 | 1.407 | 1.411 | 0 | 1 | 1 | 1 | 2 | 2 | 3 | 3 | 3 |
| 2.0 | 1.414 | 1.418 | 1.421 | 1.425 | 1.428 | 1.432 | 1.435 | 1.439 | 1.442 | 1.446 | 0 | 1 | 1 | 1 | 2 | 2 | 2 | 3 | 3 |
| 2.1 | 1.449 | 1.453 | 1.456 | 1.459 | 1.463 | 1.466 | 1.470 | 1.473 | 1.476 | 1.480 | 0 | 1 | 1 | 1 | 2 | 2 | 2 | 3 | 3 |
| 2.2 | 1.483 | 1.487 | 1.490 | 1.493 | 1.497 | 1.500 | 1.503 | 1.507 | 1.510 | 1.513 | 0 | 1 | 1 | 1 | 2 | 2 | 2 | 3 | 3 |
| 2.3 | 1.517 | 1.520 | 1.523 | 1.526 | 1.530 | 1.533 | 1.536 | 1.539 | 1.543 | 1.546 | 0 | 1 | 1 | 1 | 2 | 2 | 2 | 3 | 3 |
| 2.4 | 1.549 | 1.552 | 1.556 | 1.559 | 1.562 | 1.565 | 1.568 | 1.572 | 1.575 | 1.578 | 0 | 1 | 1 | 1 | 2 | 2 | 2 | 3 | 3 |
| 2.5 | 1.581 | 1.584 | 1.587 | 1.591 | 1.594 | 1.597 | 1.600 | 1.603 | 1.606 | 1.609 | 0 | 1 | 1 | 1 | 2 | 2 | 2 | 3 | 3 |
| 2.6 | 1.612 | 1.616 | 1.619 | 1.622 | 1.625 | 1.628 | 1.631 | 1.634 | 1.637 | 1.640 | 0 | 1 | 1 | 1 | 2 | 2 | 2 | 2 | 3 |
| 2.7 | 1.643 | 1.646 | 1.649 | 1.652 | 1.655 | 1.658 | 1.661 | 1.664 | 1.667 | 1.670 | 0 | 1 | 1 | 1 | 2 | 2 | 2 | 2 | 3 |
| 2.8 | 1.673 | 1.676 | 1.679 | 1.682 | 1.685 | 1.688 | 1.691 | 1.694 | 1.697 | 1.700 | 0 | 1 | 1 | 1 | 1 | 2 | 2 | 2 | 3 |
| 2.9 | 1.703 | 1.706 | 1.709 | 1.712 | 1.715 | 1.718 | 1.720 | 1.723 | 1.726 | 1.729 | 0 | 1 | 1 | 1 | 1 | 2 | 2 | 2 | 3 |
| 3.0 | 1.732 | 1.735 | 1.738 | 1.741 | 1.744 | 1.746 | 1.749 | 1.752 | 1.755 | 1.758 | 0 | 1 | 1 | 1 | 1 | 2 | 2 | 2 | 3 |
| 3.1 | 1.761 | 1.764 | 1.766 | 1.769 | 1.772 | 1.775 | 1.778 | 1.780 | 1.783 | 1.786 | 0 | 1 | 1 | 1 | 1 | 2 | 2 | 2 | 3 |
| 3.2 | 1.789 | 1.792 | 1.794 | 1.797 | 1.800 | 1.803 | 1.806 | 1.808 | 1.811 | 1.814 | 0 | 1 | 1 | 1 | 1 | 2 | 2 | 2 | 2 |
| 3.3 | 1.817 | 1.819 | 1.822 | 1.825 | 1.828 | 1.830 | 1.833 | 1.836 | 1.838 | 1.841 | 0 | 1 | 1 | 1 | 1 | 2 | 2 | 2 | 2 |
| 3.4 | 1.844 | 1.847 | 1.849 | 1.852 | 1.855 | 1.857 | 1.860 | 1.863 | 1.865 | 1.868 | 0 | 1 | 1 | 1 | 1 | 2 | 2 | 2 | 2 |
| 3.5 | 1.871 | 1.873 | 1.876 | 1.879 | 1.881 | 1.884 | 1.887 | 1.889 | 1.892 | 1.895 | 0 | 1 | 1 | 1 | 1 | 2 | 2 | 2 | 2 |
| 3.6 | 1.897 | 1.900 | 1.903 | 1.905 | 1.908 | 1.910 | 1.913 | 1.916 | 1.918 | 1.921 | 0 | 1 | 1 | 1 | 1 | 1 | 1 | 2 | 2 |
| 3.7 | 1.924 | 1.926 | 1.929 | 1.931 | 1.934 | 1.936 | 1.939 | 1.942 | 1.944 | 1.947 | 0 | 1 | 1 | 1 | 1 | 2 | 2 | 2 | 2 |
| 3.8 | 1.949 | 1.952 | 1.954 | 1.957 | 1.960 | 1.962 | 1.965 | 1.967 | 1.970 | 1.972 | 0 | 1 | 1 | 1 | 1 | 2 | 2 | 2 | 2 |
| 3.9 | 1.975 | 1.977 | 1.980 | 1.982 | 1.985 | 1.987 | 1.990 | 1.992 | 1.995 | 1.997 | 0 | 1 | 1 | 1 | 1 | 2 | 2 | 2 | 2 |
| 4.0 | 2.000 | 2.002 | 2.005 | 2.007 | 2.010 | 2.012 | 2.015 | 2.017 | 2.020 | 2.022 | 0 | 0 | 1 | 1 | 1 | 1 | 2 | 2 | 2 |
| 4.1 | 2.025 | 2.027 | 2.030 | 2.032 | 2.035 | 2.037 | 2.040 | 2.042 | 2.045 | 2.047 | 0 | 0 | 1 | 1 | 1 | 1 | 2 | 2 | 2 |
| 4.2 | 2.049 | 2.052 | 2.054 | 2.057 | 2.059 | 2.062 | 2.064 | 2.066 | 2.069 | 2.071 | 0 | 0 | 1 | 1 | 1 | 1 | 2 | 2 | 2 |
| 4.3 | 2.074 | 2.076 | 2.078 | 2.081 | 2.083 | 2.086 | 2.088 | 2.090 | 2.093 | 2.095 | 0 | 0 | 1 | 1 | 1 | 1 | 2 | 2 | 2 |
| 4.4 | 2.098 | 2.100 | 2.102 | 2.105 | 2.107 | 2.110 | 2.112 | 2.114 | 2.117 | 2.119 | 0 | 0 | 1 | 1 | 1 | 1 | 2 | 2 | 2 |
| 4.5 | 2.121 | 2.124 | 2.126 | 2.128 | 2.131 | 2.133 | 2.135 | 2.138 | 2.140 | 2.142 | 0 | 0 | 1 | 1 | 1 | 1 | 2 | 2 | 2 |
| 4.6 | 2.145 | 2.147 | 2.149 | 2.152 | 2.154 | 2.156 | 2.159 | 2.161 | 2.163 | 2.166 | 0 | 0 | 1 | 1 | 1 | 1 | 2 | 2 | 2 |
| 4.7 | 2.168 | 2.170 | 2.173 | 2.175 | 2.177 | 2.179 | 2.182 | 2.184 | 2.186 | 2.189 | 0 | 0 | 1 | 1 | 1 | 1 | 2 | 2 | 2 |
| 4.8 | 2.191 | 2.193 | 2.195 | 2.198 | 2.200 | 2.202 | 2.205 | 2.207 | 2.209 | 2.211 | 0 | 0 | 1 | 1 | 1 | 1 | 2 | 2 | 2 |
| 4.9 | 2.214 | 2.216 | 2.218 | 2.220 | 2.223 | 2.225 | 2.227 | 2.229 | 2.232 | 2.234 | 0 | 0 | 1 | 1 | 1 | 1 | 2 | 2 | 2 |
| 5.0 | 2.236 | 2.238 | 2.241 | 2.243 | 2.245 | 2.247 | 2.249 | 2.252 | 2.254 | 2.256 | 0 | 0 | 1 | 1 | 1 | 1 | 2 | 2 | 2 |
| 5.1 | 2.258 | 2.261 | 2.263 | 2.265 | 2.267 | 2.269 | 2.272 | 2.274 | 2.276 | 2.278 | 0 | 0 | 1 | 1 | 1 | 1 | 2 | 2 | 2 |
| 5.2 | 2.280 | 2.283 | 2.285 | 2.287 | 2.289 | 2.291 | 2.293 | 2.296 | 2.298 | 2.300 | 0 | 0 | 1 | 1 | 1 | 1 | 2 | 2 | 2 |
| 5.3 | 2.302 | 2.304 | 2.307 | 2.309 | 2.311 | 2.313 | 2.315 | 2.317 | 2.319 | 2.322 | 0 | 0 | 1 | 1 | 1 | 1 | 2 | 2 | 2 |
| 5.4 | 2.324 | 2.326 | 2.328 | 2.330 | 2.332 | 2.335 | 2.337 | 2.339 | 2.341 | 2.343 | 0 | 0 | 1 | 1 | 1 | 1 | 1 | 2 | 2 |

## TABLE D1    Square Roots (1.000 − 9.999)

| N | 0 | 1 | 2 | 3 | 4 | 5 | 6 | 7 | 8 | 9 | 1 | 2 | 3 | 4 | 5 | 6 | 7 | 8 | 9 |
|---|---|---|---|---|---|---|---|---|---|---|---|---|---|---|---|---|---|---|---|
| 5.5 | 2.345 | 2.347 | 2.349 | 2.352 | 2.354 | 2.356 | 2.358 | 2.360 | 2.362 | 2.364 | 0 | 0 | 1 | 1 | 1 | 1 | 1 | 2 | 2 |
| 5.6 | 2.366 | 2.369 | 2.371 | 2.373 | 2.375 | 2.377 | 2.379 | 2.381 | 2.383 | 2.385 | 0 | 0 | 1 | 1 | 1 | 1 | 1 | 2 | 2 |
| 5.7 | 2.387 | 2.390 | 2.392 | 2.394 | 2.396 | 2.398 | 2.400 | 2.402 | 2.404 | 2.406 | 0 | 0 | 1 | 1 | 1 | 1 | 1 | 2 | 2 |
| 5.8 | 2.408 | 2.410 | 2.412 | 2.415 | 2.417 | 2.419 | 2.421 | 2.423 | 2.425 | 2.427 | 0 | 0 | 1 | 1 | 1 | 1 | 1 | 2 | 2 |
| 5.9 | 2.429 | 2.431 | 2.433 | 2.435 | 2.437 | 2.439 | 2.441 | 2.443 | 2.445 | 2.447 | 0 | 0 | 1 | 1 | 1 | 1 | 1 | 2 | 2 |
| 6.0 | 2.449 | 2.452 | 2.454 | 2.456 | 2.458 | 2.460 | 2.462 | 2.464 | 2.466 | 2.468 | 0 | 0 | 1 | 1 | 1 | 1 | 1 | 2 | 2 |
| 6.1 | 2.470 | 2.472 | 2.474 | 2.476 | 2.478 | 2.480 | 2.482 | 2.484 | 2.486 | 2.488 | 0 | 0 | 1 | 1 | 1 | 1 | 1 | 2 | 2 |
| 6.2 | 2.490 | 2.492 | 2.494 | 2.496 | 2.498 | 2.500 | 2.502 | 2.504 | 2.506 | 2.508 | 0 | 0 | 1 | 1 | 1 | 1 | 1 | 2 | 2 |
| 6.3 | 2.510 | 2.512 | 2.514 | 2.516 | 2.518 | 2.520 | 2.522 | 2.524 | 2.526 | 2.528 | 0 | 0 | 1 | 1 | 1 | 1 | 1 | 2 | 2 |
| 6.4 | 2.530 | 2.532 | 2.534 | 2.536 | 2.538 | 2.540 | 2.542 | 2.544 | 2.546 | 2.548 | 0 | 0 | 1 | 1 | 1 | 1 | 1 | 2 | 2 |
| 6.5 | 2.550 | 2.551 | 2.553 | 2.555 | 2.557 | 2.559 | 2.561 | 2.563 | 2.565 | 2.567 | 0 | 0 | 1 | 1 | 1 | 1 | 1 | 2 | 2 |
| 6.6 | 2.569 | 2.571 | 2.573 | 2.575 | 2.577 | 2.579 | 2.581 | 2.583 | 2.585 | 2.587 | 0 | 0 | 1 | 1 | 1 | 1 | 1 | 2 | 2 |
| 6.7 | 2.588 | 2.590 | 2.592 | 2.594 | 2.596 | 2.598 | 2.600 | 2.602 | 2.604 | 2.606 | 0 | 0 | 1 | 1 | 1 | 1 | 1 | 2 | 2 |
| 6.8 | 2.608 | 2.610 | 2.612 | 2.613 | 2.615 | 2.617 | 2.619 | 2.621 | 2.623 | 2.625 | 0 | 0 | 1 | 1 | 1 | 1 | 1 | 2 | 2 |
| 6.9 | 2.627 | 2.629 | 2.631 | 2.632 | 2.634 | 2.636 | 2.638 | 2.640 | 2.642 | 2.644 | 0 | 0 | 1 | 1 | 1 | 1 | 1 | 2 | 2 |
| 7.0 | 2.646 | 2.648 | 2.650 | 2.651 | 2.653 | 2.655 | 2.657 | 2.659 | 2.661 | 2.663 | 0 | 0 | 1 | 1 | 1 | 1 | 1 | 2 | 2 |
| 7.1 | 2.665 | 2.666 | 2.668 | 2.670 | 2.672 | 2.674 | 2.676 | 2.678 | 2.680 | 2.681 | 0 | 0 | 1 | 1 | 1 | 1 | 1 | 1 | 2 |
| 7.2 | 2.683 | 2.685 | 2.687 | 2.689 | 2.691 | 2.693 | 2.694 | 2.696 | 2.698 | 2.700 | 0 | 0 | 1 | 1 | 1 | 1 | 1 | 1 | 2 |
| 7.3 | 2.702 | 2.704 | 2.706 | 2.707 | 2.709 | 2.711 | 2.713 | 2.715 | 2.717 | 2.718 | 0 | 0 | 1 | 1 | 1 | 1 | 1 | 1 | 2 |
| 7.4 | 2.720 | 2.722 | 2.724 | 2.726 | 2.728 | 2.729 | 2.731 | 2.733 | 2.735 | 2.737 | 0 | 0 | 1 | 1 | 1 | 1 | 1 | 1 | 2 |
| 7.5 | 2.793 | 2.740 | 2.742 | 2.744 | 2.746 | 2.748 | 2.750 | 2.751 | 2.753 | 2.755 | 0 | 0 | 1 | 1 | 1 | 1 | 1 | 1 | 2 |
| 7.6 | 2.757 | 2.759 | 2.760 | 2.762 | 2.764 | 2.766 | 2.768 | 2.769 | 2.771 | 2.773 | 0 | 0 | 1 | 1 | 1 | 1 | 1 | 1 | 2 |
| 7.7 | 2.775 | 2.777 | 2.778 | 2.780 | 2.782 | 2.784 | 2.786 | 2.787 | 2.789 | 2.791 | 0 | 0 | 1 | 1 | 1 | 1 | 1 | 1 | 2 |
| 7.8 | 2.793 | 2.795 | 2.796 | 2.798 | 2.800 | 2.802 | 2.804 | 2.805 | 2.807 | 2.809 | 0 | 0 | 1 | 1 | 1 | 1 | 1 | 1 | 2 |
| 7.9 | 2.811 | 2.812 | 2.814 | 2.816 | 2.818 | 2.820 | 2.821 | 2.823 | 2.825 | 2.827 | 0 | 0 | 1 | 1 | 1 | 1 | 1 | 1 | 2 |
| 8.0 | 2.828 | 2.830 | 2.832 | 2.834 | 2.835 | 2.837 | 2.839 | 2.841 | 2.843 | 2.844 | 0 | 0 | 1 | 1 | 1 | 1 | 1 | 1 | 2 |
| 8.1 | 2.846 | 2.848 | 2.850 | 2.851 | 2.853 | 2.855 | 2.857 | 2.858 | 2.860 | 2.862 | 0 | 0 | 1 | 1 | 1 | 1 | 1 | 1 | 2 |
| 8.2 | 2.864 | 2.865 | 2.867 | 2.869 | 2.871 | 2.872 | 2.874 | 2.876 | 2.877 | 2.879 | 0 | 0 | 1 | 1 | 1 | 1 | 1 | 1 | 2 |
| 8.3 | 2.881 | 2.883 | 2.884 | 2.886 | 2.888 | 2.890 | 2.891 | 2.893 | 2.895 | 2.897 | 0 | 0 | 1 | 1 | 1 | 1 | 1 | 1 | 2 |
| 8.4 | 2.898 | 2.900 | 2.902 | 2.903 | 2.905 | 2.907 | 2.909 | 2.910 | 2.912 | 2.914 | 0 | 0 | 1 | 1 | 1 | 1 | 1 | 1 | 2 |
| 8.5 | 2.915 | 2.917 | 2.919 | 2.921 | 2.922 | 2.924 | 2.926 | 2.927 | 2.929 | 2.931 | 0 | 0 | 1 | 1 | 1 | 1 | 1 | 1 | 2 |
| 8.6 | 2.933 | 2.934 | 2.936 | 2.938 | 2.939 | 2.941 | 2.943 | 2.944 | 2.946 | 2.948 | 0 | 0 | 1 | 1 | 1 | 1 | 1 | 1 | 2 |
| 8.7 | 2.950 | 2.951 | 2.953 | 2.955 | 2.956 | 2.958 | 2.960 | 2.961 | 2.963 | 2.965 | 0 | 0 | 1 | 1 | 1 | 1 | 1 | 1 | 2 |
| 8.8 | 2.966 | 2.968 | 2.970 | 2.972 | 2.973 | 2.975 | 2.977 | 2.978 | 2.980 | 2.982 | 0 | 0 | 1 | 1 | 1 | 1 | 1 | 1 | 2 |
| 8.9 | 2.983 | 2.985 | 2.987 | 2.988 | 2.990 | 2.992 | 2.993 | 2.995 | 2.997 | 2.998 | 0 | 0 | 1 | 1 | 1 | 1 | 1 | 1 | 2 |
| 9.0 | 3.000 | 3.002 | 3.003 | 3.005 | 3.007 | 3.008 | 3.010 | 3.012 | 3.013 | 3.015 | 0 | 0 | 0 | 1 | 1 | 1 | 1 | 1 | 1 |
| 9.1 | 3.017 | 3.018 | 3.020 | 3.022 | 3.023 | 3.025 | 3.027 | 3.028 | 3.030 | 3.032 | 0 | 0 | 0 | 1 | 1 | 1 | 1 | 1 | 1 |
| 9.2 | 3.033 | 3.035 | 3.036 | 3.038 | 3.040 | 3.041 | 3.043 | 3.045 | 3.046 | 3.048 | 0 | 0 | 0 | 1 | 1 | 1 | 1 | 1 | 1 |
| 9.3 | 3.050 | 3.051 | 3.053 | 3.055 | 3.056 | 3.058 | 3.059 | 3.061 | 3.063 | 3.064 | 0 | 0 | 0 | 1 | 1 | 1 | 1 | 1 | 1 |
| 9.4 | 3.066 | 3.068 | 3.069 | 3.071 | 3.072 | 3.074 | 3.076 | 3.077 | 3.079 | 3.081 | 0 | 0 | 0 | 1 | 1 | 1 | 1 | 1 | 1 |
| 9.5 | 3.082 | 3.084 | 3.085 | 3.087 | 3.089 | 3.090 | 3.092 | 3.094 | 3.095 | 3.097 | 0 | 0 | 0 | 1 | 1 | 1 | 1 | 1 | 1 |
| 9.6 | 3.098 | 3.100 | 3.102 | 3.103 | 3.105 | 3.106 | 3.108 | 3.110 | 3.111 | 3.113 | 0 | 0 | 0 | 1 | 1 | 1 | 1 | 1 | 1 |
| 9.7 | 3.114 | 3.116 | 3.118 | 3.119 | 3.121 | 3.122 | 3.124 | 3.126 | 3.127 | 3.129 | 0 | 0 | 0 | 1 | 1 | 1 | 1 | 1 | 1 |
| 9.8 | 3.130 | 3.132 | 3.134 | 3.135 | 3.137 | 3.138 | 3.140 | 3.142 | 3.143 | 3.145 | 0 | 0 | 0 | 1 | 1 | 1 | 1 | 1 | 1 |
| 9.9 | 3.146 | 3.148 | 3.150 | 3.151 | 3.153 | 3.154 | 3.156 | 3.158 | 3.159 | 3.161 | 0 | 0 | 0 | 1 | 1 | 1 | 1 | 1 | 1 |

## TABLE D1   Square Roots (10.00 – 99.99)

| N | 0 | 1 | 2 | 3 | 4 | 5 | 6 | 7 | 8 | 9 | 1 | 2 | 3 | 4 | 5 | 6 | 7 | 8 | 9 |
|---|---|---|---|---|---|---|---|---|---|---|---|---|---|---|---|---|---|---|---|
| 10 | 3.162 | 3.178 | 3.194 | 3.209 | 3.225 | 3.240 | 3.256 | 3.271 | 3.286 | 3.302 | 2 | 3 | 5 | 6 | 8 | 9 | 11 | 12 | 14 |
| 11 | 3.317 | 3.332 | 3.347 | 3.362 | 3.376 | 3.391 | 3.406 | 3.421 | 3.435 | 3.450 | 1 | 3 | 4 | 6 | 7 | 9 | 10 | 12 | 13 |
| 12 | 3.464 | 3.479 | 3.493 | 3.507 | 3.521 | 3.536 | 3.550 | 3.564 | 3.578 | 3.592 | 1 | 3 | 4 | 6 | 7 | 8 | 10 | i1 | 13 |
| 13 | 3.606 | 3.619 | 3.633 | 3.647 | 3.661 | 3.674 | 3.688 | 3.701 | 3.715 | 3.728 | 1 | 3 | 4 | 5 | 7 | 8 | 10 | 11 | 12 |
| 14 | 3.742 | 3.755 | 3.768 | 3.782 | 3.795 | 3.808 | 3.821 | 3.834 | 3.847 | 3.860 | 1 | 3 | 4 | 5 | 7 | 8 | 9 | 11 | 12 |
| 15 | 3.873 | 3.886 | 3.899 | 3.912 | 3.924 | 3.937 | 3.950 | 3.962 | 3.975 | 3.987 | 1 | 3 | 4 | 5 | 6 | 8 | 9 | 10 | 11 |
| 16 | 4.000 | 4.012 | 4.025 | 4.037 | 4.050 | 4.062 | 4.074 | 4.087 | 4.099 | 4.111 | 1 | 2 | 4 | 5 | 6 | 7 | 9 | 10 | 11 |
| 17 | 4.123 | 4.135 | 4.147 | 4.159 | 4.171 | 4.183 | 4.195 | 4.207 | 4.219 | 4.231 | 1 | 2 | 4 | 5 | 6 | 7 | 8 | i0 | 11 |
| 18 | 4.243 | 4.254 | 4.266 | 4.278 | 4.290 | 4.301 | 4.313 | 4.324 | 4.336 | 4.347 | 1 | 2 | 3 | 5 | 6 | 7 | 8 | 9 | 10 |
| 19 | 4.359 | 4.370 | 4.382 | 4.393 | 4.405 | 4.416 | 4.427 | 4.438 | 4.450 | 4.461 | 1 | 2 | 3 | 5 | 6 | 7 | 8 | 9 | 10 |
| 20 | 4.472 | 4.483 | 4.494 | 4.506 | 4.517 | 5.528 | 4.539 | 4.550 | 4.561 | 4.572 | 1 | 2 | 3 | 4 | 6 | 7 | 8 | 9 | 10 |
| 21 | 4.583 | 4.593 | 4.604 | 4.615 | 4.626 | 4.637 | 4.648 | 4.658 | 4.669 | 4.680 | 1 | 2 | 3 | 4 | 5 | 6 | 8 | 9 | 10 |
| 22 | 4.690 | 4.701 | 4.712 | 4.722 | 4.733 | 4.743 | 4.754 | 4.764 | 4.775 | 4.785 | 1 | 2 | 3 | 4 | 5 | 6 | 7 | 8 | 9 |
| 23 | 4.796 | 4.806 | 4.817 | 4.827 | 4.837 | 4.848 | 4.858 | 4.868 | 4.879 | 4.889 | 1 | 2 | 3 | 4 | 5 | 6 | 7 | 8 | 9 |
| 24 | 4.899 | 4.909 | 4.919 | 4.930 | 4.940 | 4.950 | 4.960 | 4.970 | 4.980 | 4.990 | 1 | 2 | 3 | 4 | 5 | 6 | 7 | 8 | 9 |
| 25 | 5.000 | 5.010 | 5.020 | 5.030 | 5.040 | 5.050 | 5.060 | 5.070 | 5.079 | 5.089 | 1 | 2 | 3 | 4 | 5 | 6 | 7 | 8 | 9 |
| 26 | 5.099 | 5.109 | 5.119 | 5.128 | 5.138 | 5.148 | 5.158 | 5.167 | 5.177 | 5.187 | 1 | 2 | 3 | 4 | 5 | 6 | 7 | 8 | 9 |
| 27 | 5.196 | 5.206 | 5.215 | 5.225 | 5.235 | 5.244 | 5.254 | 5.263 | 5.273 | 5.282 | 1 | 2 | 3 | 4 | 5 | 6 | 7 | 8 | 9 |
| 28 | 5.292 | 5.301 | 5.310 | 5.320 | 5.329 | 5.339 | 5.348 | 5.357 | 5.367 | 5.376 | 1 | 2 | 3 | 4 | 5 | 6 | 7 | 7 | 8 |
| 29 | 5.385 | 5.394 | 5.404 | 5.413 | 5.422 | 5.431 | 5.441 | 5.450 | 5.459 | 5.468 | 1 | 2 | 3 | 4 | 5 | 5 | 6 | 7 | 8 |
| 30 | 5.477 | 5.486 | 5.495 | 5.505 | 5.514 | 5.523 | 5.532 | 5.541 | 5.550 | 5.559 | 1 | 2 | 3 | 4 | 4 | 5 | 6 | 7 | 8 |
| 31 | 5.568 | 5.577 | 5.586 | 5.595 | 5.604 | 5.612 | 5.621 | 5.630 | 5.639 | 5.648 | 1 | 2 | 3 | 3 | 4 | 5 | 6 | 7 | 8 |
| 32 | 5.657 | 5.666 | 5.675 | 5.683 | 5.692 | 5.701 | 5.710 | 5.718 | 5.727 | 5.736 | 1 | 2 | 3 | 3 | 4 | 5 | 6 | 7 | 8 |
| 33 | 5.745 | 5.753 | 5.762 | 5.771 | 5.779 | 5.788 | 5.797 | 5.805 | 5.814 | 5.822 | 1 | 2 | 3 | 3 | 4 | 5 | 6 | 7 | 8 |
| 34 | 5.831 | 5.840 | 5.848 | 5.857 | 5.865 | 5.874 | 5.882 | 5.891 | 5.899 | 5.908 | 1 | 2 | 3 | 3 | 4 | 5 | 6 | 7 | 8 |
| 35 | 5.916 | 5.925 | 5.933 | 5.941 | 5.950 | 5.958 | 5.967 | 5.975 | 5.983 | 5.992 | 1 | 2 | 2 | 3 | 4 | 5 | 6 | 7 | 8 |
| 36 | 6.000 | 6.008 | 6.017 | 6.025 | 6.033 | 6.042 | 6.050 | 6.058 | 6.066 | 6.075 | 1 | 2 | 2 | 3 | 4 | 5 | 6 | 7 | 7 |
| 37 | 6.083 | 6.091 | 6.099 | 6.107 | 6.116 | 6.124 | 6.132 | 6.140 | 6.148 | 6.156 | 1 | 2 | 2 | 3 | 4 | 5 | 6 | 7 | 7 |
| 38 | 6.164 | 6.173 | 6.181 | 6.189 | 6.197 | 6.205 | 6.213 | 6.221 | 6.229 | 6.237 | 1 | 2 | 2 | 3 | 4 | 5 | 6 | 6 | 7 |
| 39 | 6.245 | 6.253 | 6.261 | 6.269 | 6.277 | 6.285 | 6.293 | 6.301 | 6.309 | 6.317 | 1 | 2 | 2 | 3 | 4 | 5 | 6 | 6 | 7 |
| 40 | 6.325 | 6.332 | 6.340 | 6.348 | 6.356 | 6.364 | 6.372 | 6.380 | 6.387 | 6.395 | 1 | 2 | 2 | 3 | 4 | 5 | 6 | 6 | 7 |
| 41 | 6.403 | 6.411 | 6.419 | 6.427 | 6.434 | 6.442 | 6.450 | 6.458 | 6.465 | 6.473 | 1 | 2 | 2 | 3 | 4 | 5 | 5 | 6 | 7 |
| 42 | 6.481 | 6.488 | 6.496 | 6.504 | 6.512 | 6.519 | 6.527 | 6.535 | 6.542 | 6.550 | 1 | 2 | 2 | 3 | 4 | 5 | 5 | 6 | 7 |
| 43 | 6.557 | 6.565 | 6.573 | 6.580 | 6.588 | 6.595 | 6.603 | 6.611 | 6.618 | 6.626 | 1 | 2 | 2 | 3 | 4 | 5 | 5 | 6 | 7 |
| 44 | 6.633 | 6.641 | 6.648 | 6.656 | 6.663 | 6.671 | 6.678 | 6.686 | 6.693 | 6.701 | 1 | 2 | 2 | 3 | 4 | 5 | 5 | 6 | 7 |
| 45 | 6.708 | 6.716 | 6.723 | 6.731 | 6.738 | 6.745 | 6.753 | 6.760 | 6.768 | 6.775 | 1 | 1 | 2 | 3 | 4 | 4 | 5 | 6 | 7 |
| 46 | 6.782 | 6.790 | 6.797 | 6.804 | 6.812 | 6.819 | 6.826 | 6.834 | 6.841 | 6.848 | 1 | 1 | 2 | 3 | 4 | 4 | 5 | 6 | 7 |
| 47 | 6.856 | 6.863 | 6.870 | 6.877 | 6.885 | 6.892 | 6.899 | 6.907 | 6.914 | 6.921 | 1 | 1 | 2 | 3 | 4 | 4 | 5 | 6 | 7 |
| 48 | 6.928 | 6.935 | 6.943 | 6.950 | 6.957 | 6.964 | 6.971 | 6.979 | 6.986 | 6.993 | 1 | 1 | 2 | 3 | 4 | 4 | 5 | 6 | 6 |
| 49 | 7.000 | 7.007 | 7.014 | 7.021 | 7.029 | 7.036 | 7.043 | 7.050 | 7.057 | 7.064 | 1 | 1 | 2 | 3 | 4 | 4 | 5 | 6 | 6 |
| 50 | 7.071 | 7.078 | 7.085 | 7.092 | 7.099 | 7.106 | 7.113 | 7.120 | 7.127 | 7.134 | 1 | 1 | 2 | 3 | 4 | 4 | 5 | 6 | 6 |
| 51 | 7.141 | 7.148 | 7.155 | 7.162 | 7.169 | 7.176 | 7.183 | 7.190 | 7.197 | 7.204 | 1 | 1 | 2 | 3 | 4 | 4 | 5 | 6 | 6 |
| 52 | 7.211 | 7.218 | 7.225 | 7.232 | 7.239 | 7.246 | 7.253 | 7.259 | 7.266 | 7.273 | 1 | 1 | 2 | 3 | 3 | 4 | 5 | 6 | 6 |
| 53 | 7.280 | 7.287 | 7.294 | 7.301 | 7.308 | 7.314 | 7.321 | 7.328 | 7.335 | 7.342 | 1 | 1 | 2 | 3 | 3 | 4 | 5 | 5 | 6 |
| 54 | 7.348 | 7.355 | 7.362 | 7.369 | 7.376 | 7.382 | 7.389 | 7.396 | 7.403 | 7.409 | 1 | 1 | 2 | 3 | 3 | 4 | 5 | 5 | 6 |

### TABLE D1  Square Roots (10.00 – 99.99)

| N | 0 | 1 | 2 | 3 | 4 | 5 | 6 | 7 | 8 | 9 | 1 | 2 | 3 | 4 | 5 | 6 | 7 | 8 | 9 |
|---|---|---|---|---|---|---|---|---|---|---|---|---|---|---|---|---|---|---|---|
| 55 | 7.416 | 7.423 | 7.430 | 7.436 | 7.443 | 7.450 | 7.457 | 7.463 | 7.470 | 7.477 | 1 | 1 | 2 | 3 | 3 | 4 | 5 | 5 | 6 |
| 56 | 7.483 | 7.490 | 7.497 | 7.503 | 7.510 | 7.517 | 7.523 | 7.530 | 7.537 | 7.543 | 1 | 1 | 2 | 3 | 3 | 4 | 5 | 5 | 6 |
| 57 | 7.550 | 7.556 | 7.563 | 7.570 | 7.576 | 7.583 | 7.589 | 7.596 | 7.603 | 7.609 | 1 | 1 | 2 | 3 | 3 | 4 | 5 | 5 | 6 |
| 58 | 7.616 | 7.622 | 7.629 | 7.635 | 7.642 | 7.649 | 7.655 | 7.662 | 7.668 | 7.675 | 1 | 1 | 2 | 3 | 3 | 4 | 5 | 5 | 6 |
| 59 | 7.681 | 7.688 | 7.694 | 7.701 | 7.707 | 7.714 | 7.720 | 7.727 | 7.733 | 7.740 | 1 | 1 | 2 | 3 | 3 | 4 | 4 | 5 | 6 |
| 60 | 7.746 | 7.752 | 7.759 | 7.765 | 7.772 | 7.778 | 7.785 | 7.791 | 7.797 | 7.804 | 1 | 1 | 2 | 3 | 3 | 4 | 4 | 5 | 6 |
| 61 | 7.810 | 7.817 | 7.823 | 7.829 | 7.836 | 7.842 | 7.849 | 7.855 | 7.861 | 7.868 | 1 | 1 | 2 | 3 | 3 | 4 | 4 | 5 | 6 |
| 62 | 7.874 | 7.880 | 7.887 | 7.893 | 7.899 | 7.906 | 7.912 | 7.918 | 7.925 | 7.931 | 1 | 1 | 2 | 3 | 3 | 4 | 4 | 5 | 6 |
| 63 | 7.937 | 7.944 | 7.950 | 7.956 | 7.962 | 7.969 | 7.975 | 7.981 | 7.987 | 7.994 | 1 | 1 | 2 | 3 | 3 | 4 | 4 | 5 | 6 |
| 64 | 8.000 | 8.006 | 8.012 | 8.019 | 8.025 | 8.031 | 8.037 | 8.044 | 8.050 | 8.056 | 1 | 1 | 2 | 2 | 3 | 4 | 4 | 5 | 6 |
| 65 | 8.062 | 8.068 | 8.075 | 8.081 | 8.087 | 8.093 | 8.099 | 8.106 | 8.112 | 8.118 | 1 | 1 | 2 | 2 | 3 | 4 | 4 | 5 | 5 |
| 66 | 8.124 | 8.130 | 8.136 | 8.142 | 8.149 | 8.155 | 8.161 | 8.167 | 8.173 | 8.179 | 1 | 1 | 2 | 2 | 3 | 4 | 4 | 5 | 5 |
| 67 | 8.185 | 8.191 | 8.198 | 8.204 | 8.210 | 8.216 | 8.222 | 8.228 | 8.234 | 8.240 | 1 | 1 | 2 | 2 | 3 | 4 | 4 | 5 | 5 |
| 68 | 8.246 | 8.252 | 8.258 | 8.264 | 8.270 | 8.276 | 8.283 | 8.289 | 8.295 | 8.301 | 1 | 1 | 2 | 2 | 3 | 4 | 4 | 5 | 5 |
| 69 | 8.307 | 8.313 | 8.319 | 8.325 | 8.331 | 8.337 | 8.343 | 8.340 | 8.355 | 8.361 | 1 | 1 | 2 | 2 | 3 | 4 | 4 | 5 | 5 |
| 70 | 8.367 | 8.373 | 8.379 | 8.385 | 8.390 | 8.396 | 8.402 | 8.408 | 8.414 | 8.420 | 1 | 1 | 2 | 2 | 3 | 4 | 4 | 5 | 5 |
| 71 | 8.426 | 8.432 | 8.438 | 8.444 | 8.450 | 8.456 | 8.462 | 8.468 | 8.473 | 8.479 | 1 | 1 | 2 | 2 | 3 | 4 | 4 | 5 | 5 |
| 72 | 8.485 | 8.491 | 8.497 | 8.503 | 8.509 | 8.515 | 8.521 | 8.526 | 8.532 | 8.538 | 1 | 1 | 2 | 2 | 3 | 3 | 4 | 5 | 5 |
| 73 | 8.544 | 8.550 | 8.556 | 8.562 | 8.567 | 8.573 | 8.579 | 8.585 | 8.591 | 8.597 | 1 | 1 | 2 | 2 | 3 | 3 | 4 | 5 | 5 |
| 74 | 8.602 | 8.608 | 8.614 | 8.620 | 8.626 | 8.631 | 8.637 | 8.643 | 8.649 | 8.654 | 1 | 1 | 2 | 2 | 3 | 3 | 4 | 5 | 5 |
| 75 | 8.660 | 8.666 | 8.672 | 8.678 | 8.683 | 8.689 | 8.695 | 8.701 | 8.706 | 8.712 | 1 | 1 | 2 | 2 | 3 | 3 | 4 | 5 | 5 |
| 76 | 8.718 | 8.724 | 8.729 | 8.735 | 8.741 | 8.746 | 8.752 | 8.758 | 8.764 | 8.769 | 1 | 1 | 2 | 2 | 3 | 3 | 4 | 5 | 5 |
| 77 | 8.775 | 8.781 | 8.786 | 8.792 | 8.798 | 8.803 | 8.809 | 8.815 | 8.820 | 8.826 | 1 | 1 | 2 | 2 | 3 | 3 | 4 | 4 | 5 |
| 78 | 8.832 | 8.837 | 8.843 | 8.849 | 8.854 | 8.860 | 8.866 | 8.871 | 8.877 | 8.883 | 1 | 1 | 2 | 2 | 3 | 3 | 4 | 4 | 5 |
| 79 | 8.888 | 8.894 | 8.899 | 8.905 | 8.911 | 8.916 | 8.922 | 8.927 | 8.933 | 8.939 | 1 | 1 | 2 | 2 | 3 | 3 | 4 | 4 | 5 |
| 80 | 8.944 | 8.950 | 8.955 | 8.961 | 8.967 | 8.972 | 8.978 | 8.983 | 8.989 | 8.994 | 1 | 1 | 2 | 2 | 3 | 3 | 4 | 4 | 5 |
| 81 | 9.000 | 9.006 | 9.011 | 9.017 | 9.022 | 9.028 | 9.033 | 9.039 | 9.044 | 9.050 | 1 | 1 | 2 | 2 | 3 | 3 | 4 | 4 | 5 |
| 82 | 9.055 | 9.061 | 9.066 | 9.072 | 9.077 | 9.083 | 9.088 | 9.094 | 9.099 | 9.105 | 1 | 1 | 2 | 2 | 3 | 3 | 4 | 4 | 5 |
| 83 | 9.110 | 9.116 | 9.121 | 9.127 | 9.132 | 9.138 | 9.143 | 9.149 | 9.154 | 9.160 | 1 | 1 | 2 | 2 | 3 | 3 | 4 | 4 | 5 |
| 84 | 9.165 | 9.171 | 9.176 | 9.182 | 9.187 | 9.192 | 9.198 | 9.203 | 9.209 | 9.214 | 1 | 1 | 2 | 2 | 3 | 3 | 4 | 4 | 5 |
| 85 | 9.220 | 9.225 | 9.230 | 9.236 | 9.241 | 9.247 | 9.252 | 9.257 | 9.263 | 9.268 | 1 | 1 | 2 | 2 | 3 | 3 | 4 | 4 | 5 |
| 86 | 9.274 | 9.279 | 9.284 | 9.290 | 9.295 | 9.301 | 9.306 | 9.311 | 9.317 | 9.322 | 1 | 1 | 2 | 2 | 3 | 3 | 4 | 4 | 5 |
| 87 | 9.327 | 9.333 | 9.338 | 9.343 | 9.349 | 9.354 | 9.359 | 9.365 | 9.370 | 9.375 | 1 | 1 | 2 | 2 | 3 | 3 | 4 | 4 | 5 |
| 88 | 9.381 | 9.386 | 9.391 | 9.397 | 9.402 | 9.407 | 9.413 | 9.418 | 9.423 | 9.429 | 1 | 1 | 2 | 2 | 3 | 3 | 4 | 4 | 5 |
| 89 | 9.434 | 9.439 | 9.445 | 9.450 | 9.455 | 9.460 | 9.466 | 9.471 | 9.476 | 9.482 | 1 | 1 | 2 | 2 | 3 | 3 | 4 | 4 | 5 |
| 90 | 9.487 | 9.492 | 9.497 | 9.503 | 9.508 | 9.513 | 9.518 | 9.524 | 9.529 | 9.534 | 1 | 1 | 2 | 2 | 3 | 3 | 4 | 4 | 5 |
| 91 | 9.539 | 9.545 | 9.550 | 9.555 | 9.560 | 9.566 | 9.571 | 9.576 | 9.581 | 9.586 | 1 | 1 | 2 | 2 | 3 | 3 | 4 | 4 | 5 |
| 92 | 9.592 | 9.597 | 9.602 | 9.607 | 9.612 | 9.618 | 9.623 | 9.628 | 9.633 | 9.638 | 1 | 1 | 2 | 2 | 3 | 3 | 4 | 4 | 5 |
| 93 | 9.644 | 9.649 | 9.654 | 9.659 | 9.664 | 9.670 | 9.675 | 9.680 | 9.685 | 9.690 | 1 | 1 | 2 | 2 | 3 | 3 | 4 | 4 | 5 |
| 94 | 9.695 | 9.701 | 9.706 | 9.711 | 9.716 | 9.721 | 9.726 | 9.731 | 9.737 | 9.742 | 1 | 1 | 2 | 2 | 3 | 3 | 4 | 4 | 5 |
| 95 | 9.747 | 9.752 | 9.757 | 9.762 | 9.767 | 9.772 | 9.778 | 9.783 | 9.788 | 9.793 | 1 | 1 | 2 | 2 | 3 | 3 | 4 | 4 | 5 |
| 96 | 0.798 | 9.803 | 9.808 | 9.813 | 9.818 | 9.823 | 9.829 | 9.834 | 9.839 | 9.844 | 1 | 1 | 2 | 2 | 3 | 3 | 4 | 4 | 5 |
| 97 | 9.849 | 9.854 | 9.859 | 9.864 | 9.869 | 9.874 | 9.879 | 9.884 | 9.889 | 9.894 | 1 | 1 | 2 | 2 | 3 | 3 | 4 | 4 | 5 |
| 98 | 9.899 | 9.905 | 9.910 | 9.915 | 9.920 | 9.925 | 9.930 | 9.935 | 9.940 | 9.945 | 1 | 1 | 1 | 2 | 2 | 3 | 3 | 4 | 4 |
| 99 | 9.950 | 9.955 | 9.960 | 9.965 | 9.970 | 9.975 | 9.980 | 9.985 | 9.990 | 9.995 | 0 | 1 | 1 | 2 | 2 | 3 | 3 | 4 | 4 |

**TABLE D2    Trigonometric Values, Angle θ in Degrees and Tenths**

| Degrees | Sin θ | Cos θ | Tan θ | Cot θ | | Degrees | Sin θ | Cos θ | Tan θ | Cot θ | |
|---|---|---|---|---|---|---|---|---|---|---|---|
| 0.0 | 0.0000 | 1.0000 | 0.0000 | – | 90.0 | 5.0 | 0.0872 | 0.9962 | 0.0875 | 11.43 | 85.0 |
| .1 | 0.0017 | 1.0000 | 0.0017 | 573.0 | .9 | .1 | 0.0889 | 0.9960 | 0.0892 | 11.20 | .9 |
| .2 | 0.0035 | 1.0000 | 0.0035 | 286.4 | .8 | .2 | 0.0906 | 0.9959 | 0.0910 | 10.99 | .8 |
| .3 | 0.0052 | 1.0000 | 0.0052 | 191.0 | .7 | .3 | 0.0924 | 0.9957 | 0.0928 | 10.78 | .7 |
| .4 | 0.0070 | 1.0000 | 0.0070 | 143.2 | .6 | .4 | 0.0941 | 0.9956 | 0.0945 | 10.58 | .6 |
| .5 | 0.0087 | 1.0000 | 0.0087 | 114.6 | .5 | .5 | 0.0958 | 0.9954 | 0.0963 | 10.39 | .5 |
| .6 | 0.0105 | 0.9999 | 0.0105 | 95.49 | .4 | .6 | 0.0976 | 0.9952 | 0.0981 | 10.20 | .4 |
| .7 | 0.0122 | 0.9999 | 0.0122 | 81.85 | .3 | .7 | 0.0993 | 0.9951 | 0.0998 | 10.02 | .3 |
| .8 | 0.0140 | 0.9999 | 0.0140 | 71.62 | .2 | .8 | 0.1011 | 0.9949 | 0.1016 | 9.845 | .2 |
| .9 | 0.0157 | 0.9999 | 0.0157 | 63.66 | .1 | .9 | 0.1028 | 0.9947 | 0.1033 | 9.677 | .1 |
| 1.0 | 0.0175 | 0.9998 | 0.0175 | 57.29 | 89.0 | 6.0 | 0.1045 | 0.9945 | 0.1051 | 9.514 | 84.0 |
| .1 | 0.0192 | 0.9998 | 0.0192 | 52.08 | .9 | .1 | 0.1063 | 0.9943 | 0.1069 | 9.357 | .9 |
| .2 | 0.0209 | 0.9998 | 0.0209 | 47.74 | .8 | .2 | 0.1080 | 0.9942 | 0.1086 | 9.205 | .8 |
| .3 | 0.0227 | 0.9997 | 0.0227 | 44.07 | .7 | .3 | 0.1097 | 0.9940 | 0.1104 | 9.058 | .7 |
| .4 | 0.0244 | 0.9997 | 0.0244 | 40.92 | .6 | .4 | 0.1115 | 0.9938 | 0.1122 | 8.915 | .6 |
| .5 | 0.0262 | 0.9997 | 0.0262 | 38.19 | .5 | .5 | 0.1132 | 0.9936 | 0.1139 | 8.777 | .5 |
| .6 | 0.0279 | 0.9996 | 0.0279 | 35.80 | .4 | .6 | 0.1149 | 0.9934 | 0.1157 | 8.643 | .4 |
| .7 | 0.0297 | 0.9996 | 0.0297 | 33.69 | .3 | .7 | 0.1167 | 0.9932 | 0.1175 | 8.513 | .3 |
| .8 | 0.0314 | 0.9995 | 0.0314 | 31.82 | .2 | .8 | 0.1184 | 0.9930 | 0.1192 | 8.386 | .2 |
| .9 | 0.0332 | 0.9995 | 0.0332 | 30.14 | .1 | .9 | 0.1201 | 0.9928 | 0.1210 | 8.264 | .1 |
| 2.0 | 0.0349 | 0.9994 | 0.0349 | 28.64 | 88.0 | 7.0 | 0.1219 | 0.9925 | 0.1228 | 8.144 | 83.0 |
| .1 | 0.0366 | 0.9993 | 0.0367 | 27.27 | .9 | .1 | 0.1236 | 0.9923 | 0.1246 | 8.028 | .9 |
| .2 | 0.0384 | 0.9993 | 0.0384 | 26.03 | .8 | .2 | 0.1253 | 0.9921 | 0.1263 | 7.916 | .8 |
| .3 | 0.0401 | 0.9992 | 0.0402 | 24.90 | .7 | .3 | 0.1271 | 0.9919 | 0.1281 | 7.806 | .7 |
| .4 | 0.0419 | 0.9991 | 0.0419 | 23.86 | .6 | .4 | 0.1288 | 0.9917 | 0.1299 | 7.700 | .6 |
| .5 | 0.0436 | 0.9990 | 0.0437 | 22.90 | .5 | .5 | 0.1305 | 0.9914 | 0.1317 | 7.596 | .5 |
| .6 | 0.0454 | 0.9990 | 0.0454 | 22.02 | .4 | .6 | 0.1323 | 0.9912 | 0.1334 | 7.495 | .4 |
| .7 | 0.0471 | 0.9989 | 0.0472 | 21.20 | .3 | .7 | 0.1340 | 0.9910 | 0.1352 | 7.396 | .3 |
| .8 | 0.0488 | 0.9988 | 0.0489 | 20.45 | .2 | .8 | 0.1357 | 0.9907 | 0.1370 | 7.300 | .2 |
| .9 | 0.0506 | 0.9987 | 0.0507 | 19.74 | .1 | .9 | 0.1374 | 0.9905 | 0.1388 | 7.207 | .1 |
| 3.0 | 0.0523 | 0.9986 | 0.0524 | 19.08 | 87.0 | 8.0 | 0.1392 | 0.9903 | 0.1405 | 7.115 | 82.0 |
| .1 | 0.0541 | 0.9985 | 0.0542 | 18.46 | .9 | .1 | 0.1409 | 0.9900 | 0.1423 | 7.026 | 9 |
| .2 | 0.0558 | 0.9984 | 0.0559 | 17.89 | .8 | .2 | 0.1426 | 0.9898 | 0.1441 | 6.940 | .8 |
| .3 | 0.0576 | 0.9983 | 0.0577 | 17.34 | .7 | .3 | 0.1444 | 0.9895 | 0.1459 | 6.855 | .7 |
| .4 | 0.0593 | 0.9982 | 0.0594 | 16.83 | .6 | .4 | 0.1461 | 0.9893 | 0.1477 | 6.772 | .6 |
| .5 | 0.0610 | 0.9981 | 0.0612 | 16.35 | .5 | .5 | 0.1478 | 0.9890 | 0.1495 | 6.691 | .5 |
| .6 | 0.0628 | 0.9980 | 0.0629 | 15.89 | .4 | .6 | 0.1495 | 0.9888 | 0.1512 | 6.612 | .4 |
| .7 | 0.0645 | 0.9979 | 0.0647 | 15.46 | .3 | .7 | 0.1513 | 0.9885 | 0.1530 | 6.535 | .3 |
| .8 | 0.0663 | 0.9978 | 0.0664 | 15.06 | .2 | .8 | 0.1530 | 0.9882 | 0.1548 | 6.460 | .2 |
| .9 | 0.0680 | 0.9977 | 0.0682 | 14.67 | .1 | .9 | 0.1547 | 0.9880 | 0.1566 | 6.386 | .1 |
| 4.0 | 0.0698 | 0.9976 | 0.0699 | 14.30 | 86.0 | 9.0 | 0.1564 | 0.9877 | 0.1584 | 6.314 | 81.0 |
| .1 | 0.0715 | 0.9974 | 0.0717 | 13.95 | .9 | .1 | 0.1582 | 0.9874 | 0.1602 | 6.243 | .9 |
| .2 | 0.0732 | 0.9973 | 0.0734 | 13.62 | .8 | .2 | 0.1599 | 0.9871 | 0.1620 | 6.174 | .8 |
| .3 | 0.0750 | 0.9972 | 0.0752 | 13.30 | .7 | .3 | 0.1616 | 0.9869 | 0.1638 | 6.107 | .7 |
| .4 | 0.0767 | 0.9971 | 0.0769 | 13.00 | .6 | .4 | 0.1633 | 0.9866 | 0.1655 | 6.041 | .6 |
| .5 | 0.0785 | 0.9969 | 0.0787 | 12.71 | .5 | .5 | 0.1650 | 0.9863 | 0.1673 | 5.976 | .5 |
| .6 | 0.0802 | 0.9968 | 0.0805 | 12.43 | .4 | .6 | 0.1668 | 0.9860 | 0.1691 | 5.912 | .4 |
| .7 | 0.0819 | 0.9966 | 0.0822 | 12.16 | .3 | .7 | 0.1685 | 0.9857 | 0.1709 | 5.850 | .3 |
| .8 | 0.0837 | 0.9965 | 0.0840 | 11.91 | .2 | .8 | 0.1702 | 0.9854 | 0.1727 | 5.789 | .2 |
| .9 | 0.0854 | 0.9963 | 0.0857 | 11.66 | .1 | .9 | 0.1719 | 0.9851 | 0.1745 | 5.730 | .1 |
| 5.0 | 0.0872 | 0.9962 | 0.0875 | 11.43 | 85.0 | 10.0 | 0.1736 | 0.9848 | 0.1763 | 5.671 | 80.0 |
| | Cos θ | Sin θ | Cot θ | Tan θ | Degrees | | Cos θ | Sin θ | Cot θ | Tan θ | Degrees |

**TABLE D2   Trigonometric Values, Angle θ in Degrees and Tenths**

| Degrees | Sin θ | Cos θ | Tan θ | Cot θ | | Degrees | Sin θ | Cos θ | Tan θ | Cot θ | |
|---------|-------|-------|-------|-------|------|---------|-------|-------|-------|-------|------|
| 10.0 | 0.1736 | 0.9848 | 0.1763 | 5.671 | 80.0 | 15.0 | 0.2588 | 0.9659 | 0.2679 | 3.732 | 75.0 |
| .1 | 0.1754 | 0.9845 | 0.1781 | 5.614 | .9 | .1 | 0.2605 | 0.9655 | 0.2698 | 3.706 | .9 |
| .2 | 0.1771 | 0.9842 | 0.1799 | 5.558 | .8 | .2 | 0.2622 | 0.9650 | 0.2717 | 3.681 | .8 |
| .3 | 0.1788 | 0.9839 | 0.1817 | 5.503 | .7 | .3 | 0.2639 | 0.9646 | 0.2736 | 3.655 | .7 |
| .4 | 0.1805 | 0.9836 | 0.1835 | 5.449 | .6 | .4 | 0.2656 | 0.9641 | 0.2754 | 3.630 | .6 |
| .5 | 0.1822 | 0.9833 | 0.1853 | 5.396 | .5 | .5 | 0.2672 | 0.9636 | 0.2773 | 3.606 | .5 |
| .6 | 0.1840 | 0.9829 | 0.1871 | 5.343 | .4 | .6 | 0.2689 | 0.9632 | 0.2792 | 3.582 | .4 |
| .7 | 0.1857 | 0.9826 | 0.1890 | 5.292 | .3 | .7 | 0.2706 | 0.9627 | 0.2811 | 3.558 | .3 |
| .8 | 0.1874 | 0.9823 | 0.1908 | 5.242 | .2 | .8 | 0.2723 | 0.9622 | 0.2830 | 3.534 | .2 |
| .9 | 0.1891 | 0.9820 | 0.1926 | 5.193 | .1 | .9 | 0.2740 | 0.9617 | 0.2849 | 3.511 | .1 |
| 11.0 | 0.1908 | 0.9816 | 0.1944 | 5.145 | 79.0 | 16.0 | 0.2756 | 0.9613 | 0.2867 | 3.487 | 74.0 |
| .1 | 0.1925 | 0.9813 | 0.1962 | 5.097 | .9 | .1 | 0.2773 | 0.9608 | 0.2886 | 3.465 | .9 |
| .2 | 0.1942 | 0.9810 | 0.1980 | 5.050 | .8 | .2 | 0.2790 | 0.9603 | 0.2905 | 3.442 | .8 |
| .3 | 0.1959 | 0.9806 | 0.1998 | 5.005 | .7 | .3 | 0.2807 | 0.9598 | 0.2924 | 3.420 | .7 |
| .4 | 0.1977 | 0.9803 | 0.2016 | 4.959 | .6 | .4 | 0.2823 | 0.9593 | 0.2943 | 3.398 | .6 |
| .5 | 0.1994 | 0.9799 | 0.2035 | 4.915 | .5 | .5 | 0.2840 | 0.9588 | 0.2962 | 3.376 | .5 |
| .6 | 0.2011 | 0.9796 | 0.2053 | 4.872 | .4 | .6 | 0.2857 | 0.9583 | 0.2981 | 3.354 | .4 |
| .7 | 0.2028 | 0.9792 | 0.2071 | 4.829 | .3 | .7 | 0.2874 | 0.9578 | 0.3000 | 3.333 | .3 |
| .8 | 0.2045 | 0.9789 | 0.2089 | 4.787 | .2 | .8 | 0.2890 | 0.9573 | 0.3019 | 3.312 | .2 |
| .9 | 0.2062 | 0.9785 | 0.2107 | 4.745 | .1 | .9 | 0.2907 | 0.9568 | 0.3038 | 3.291 | .1 |
| 12.0 | 0.2079 | 0.9781 | 0.2126 | 4.705 | 78.0 | 17.0 | 0.2924 | 0.9563 | 0.3057 | 3.271 | 73.0 |
| .1 | 0.2096 | 0.9778 | 0.2144 | 4.665 | .9 | .1 | 0.2940 | 0.9558 | 0.3076 | 3.251 | .9 |
| .2 | 0.2113 | 0.9774 | 0.2162 | 4.625 | .8 | .2 | 0.2957 | 0.9553 | 0.3096 | 3.230 | .8 |
| .3 | 0.2130 | 0.9770 | 0.2180 | 4.586 | .7 | .3 | 0.2974 | 0.9548 | 0.3115 | 3.211 | .7 |
| .4 | 0.2147 | 0.9767 | 0.2199 | 4.548 | .6 | .4 | 0.2990 | 0.9542 | 0.3134 | 3.191 | .6 |
| .5 | 0.2164 | 0.9763 | 0.2217 | 4.511 | .5 | .5 | 0.3007 | 0.9537 | 0.3153 | 3.172 | .5 |
| .6 | 0.2181 | 0.9759 | 0.2235 | 4.474 | .4 | .6 | 0.3024 | 0.9532 | 0.3172 | 3.152 | .4 |
| .7 | 0.2198 | 0.9755 | 0.2254 | 4.437 | .3 | .7 | 0.3040 | 0.9527 | 0.3191 | 3.133 | .3 |
| .8 | 0.2215 | 0.9751 | 0.2272 | 4.402 | .2 | .8 | 0.3057 | 0.9521 | 0.3211 | 3.115 | .2 |
| .9 | 0.2233 | 0.9748 | 0.2290 | 4.366 | .1 | .9 | 0.3074 | 0.9516 | 0.3230 | 3.096 | .1 |
| 13.0 | 0.2250 | 0.9744 | 0.2309 | 4.331 | 77.0 | 18.0 | 0.3090 | 0.9511 | 0.3249 | 3.078 | 72.0 |
| .1 | 0.2267 | 0.9740 | 0.2327 | 4.297 | .9 | .1 | 0.3107 | 0.9505 | 0.3269 | 3.060 | .9 |
| .2 | 0.2284 | 0.9736 | 0.2345 | 4.264 | .8 | .2 | 0.3123 | 0.9500 | 0.3288 | 3.042 | .8 |
| .3 | 0.2300 | 0.9732 | 0.2364 | 4.230 | .7 | .3 | 0.3140 | 0.9494 | 0.3307 | 3.024 | .7 |
| .4 | 0.2317 | 0.9728 | 0.2382 | 4.198 | .6 | .4 | 0.3156 | 0.9489 | 0.3327 | 3.006 | .6 |
| .5 | 0.2334 | 0.9724 | 0.2401 | 4.165 | .5 | .5 | 0.3173 | 0.9483 | 0.3346 | 2.989 | .5 |
| .6 | 0.2351 | 0.9720 | 0.2419 | 4.134 | .4 | .6 | 0.3190 | 0.9478 | 0.3365 | 2.971 | .4 |
| .7 | 0.2368 | 0.9715 | 0.2438 | 4.102 | .3 | .7 | 0.3206 | 0.9472 | 0.3385 | 2.954 | .3 |
| .8 | 0.2385 | 0.9711 | 0.2456 | 4.071 | .2 | .8 | 0.3223 | 0.9466 | 0.3404 | 2.937 | .2 |
| .9 | 0.2402 | 0.9707 | 0.2475 | 4.041 | .1 | .9 | 0.3239 | 0.9461 | 0.3424 | 2.921 | .1 |
| 14.0 | 0.2419 | 0.9703 | 0.2493 | 4.011 | 76.0 | 19.0 | 0.3256 | 0.9455 | 0.3443 | 2.904 | 71.0 |
| .1 | 0.2436 | 0.9699 | 0.2512 | 3.981 | .9 | .1 | 0.3272 | 0.9449 | 0.3463 | 2.888 | .9 |
| .2 | 0.2453 | 0.9694 | 0.2530 | 3.952 | .8 | .2 | 0.3289 | 0.9444 | 0.3482 | 2.872 | .8 |
| .3 | 0.2470 | 0.9690 | 0.2549 | 3.923 | .7 | .3 | 0.3305 | 0.9438 | 0.3502 | 2.856 | .7 |
| .4 | 0.2487 | 0.9686 | 0.2568 | 3.895 | .6 | .4 | 0.3322 | 0.9432 | 0.3522 | 2.840 | .6 |
| .5 | 0.2504 | 0.9681 | 0.2586 | 3.867 | .5 | .5 | 0.3338 | 0.9426 | 0.3541 | 2.824 | .5 |
| .6 | 0.2521 | 0.9677 | 0.2605 | 3.839 | .4 | .6 | 0.3355 | 0.9421 | 0.3561 | 2.808 | .4 |
| .7 | 0.2538 | 0.9673 | 0.2623 | 3.812 | .3 | .7 | 0.3371 | 0.9415 | 0.3581 | 2.793 | .3 |
| .8 | 0.2554 | 0.9668 | 0.2642 | 3.785 | .2 | .8 | 0.3387 | 0.9409 | 0.3600 | 2.778 | .2 |
| .9 | 0.2571 | 0.9664 | 0.2661 | 3.758 | .1 | .9 | 0.3404 | 0.9403 | 0.3620 | 2.762 | .1 |
| 15.0 | 0.2588 | 0.9659 | 0.2679 | 3.732 | 75.0 | 20.0 | 0.3420 | 0.9397 | 0.3640 | 2.747 | 70.0 |
| | Cos θ | Sin θ | Cot θ | Tan θ | Degrees | | Cos θ | Sin θ | Cot θ | Tan θ | Degrees |

## TABLE D2   Trigonometric Values, Angle θ in Degrees and Tenths

| Degrees | Sin θ | Cos θ | Tan θ | Cot θ | | Degrees | Sin θ | Cos θ | Tan θ | Cot θ | |
|---------|-------|-------|-------|-------|---|---------|-------|-------|-------|-------|---|
| 20.0 | 0.3420 | 0.9397 | 0.3640 | 2.747 | 70.0 | 25.0 | 0.4226 | 0.9063 | 0.4663 | 2.145 | 65.0 |
| .1 | 0.3437 | 0.9391 | 0.3659 | 2.733 | .9 | .1 | 0.4242 | 0.9056 | 0.4684 | 2.135 | .9 |
| .2 | 0.3453 | 0.9385 | 0.3679 | 2.718 | .8 | .2 | 0.4258 | 0.9048 | 0.4706 | 2.125 | .8 |
| .3 | 0.3469 | 0.9379 | 0.3699 | 2.703 | .7 | .3 | 0.4274 | 0.9041 | 0.4727 | 2.116 | .7 |
| .4 | 0.3486 | 0.9373 | 0.3719 | 2.689 | .6 | .4 | 0.4289 | 0.9033 | 0.4748 | 2.106 | .6 |
| .5 | 0.3502 | 0.9367 | 0.3739 | 2.675 | .5 | .5 | 0.4305 | 0.9026 | 0.4770 | 2.097 | .5 |
| .6 | 0.3518 | 0.9361 | 0.3759 | 2.660 | .4 | .6 | 0.4321 | 0.9018 | 0.4791 | 2.087 | .4 |
| .7 | 0.3535 | 0.9354 | 0.3779 | 2.646 | .3 | .7 | 0.4337 | 0.9011 | 0.4813 | 2.078 | .3 |
| .8 | 0.3551 | 0.9348 | 0.3799 | 2.633 | .2 | .8 | 0.4352 | 0.9003 | 0.4834 | 2.069 | .2 |
| .9 | 0.3567 | 0.9342 | 0.3819 | 2.619 | .1 | .9 | 0.4368 | 0.8996 | 0.4856 | 2.059 | .1 |
| 21.0 | 0.3584 | 0.9336 | 0.3839 | 2.605 | 69.0 | 26.0 | 0.4384 | 0.8988 | 0.4877 | 2.050 | 64.0 |
| .1 | 0.3600 | 0.9330 | 0.3859 | 2.592 | .9 | .1 | 0.4399 | 0.8980 | 0.4899 | 2.041 | .9 |
| .2 | 0.3616 | 0.9323 | 0.3879 | 2.578 | .8 | .2 | 0.4415 | 0.8973 | 0.4921 | 2.032 | .8 |
| .3 | 0.3633 | 0.9317 | 0.3899 | 2.565 | .7 | .3 | 0.4431 | 0.8965 | 0.4942 | 2.023 | .7 |
| .4 | 0.3649 | 0.9311 | 0.3919 | 2.552 | .6 | .4 | 0.4446 | 0.8957 | 0.4964 | 2.014 | .6 |
| .5 | 0.3665 | 0.9304 | 0.3939 | 2.539 | .5 | .5 | 0.4462 | 0.8949 | 0.4986 | 2.006 | .5 |
| .6 | 0.3681 | 0.9298 | 0.3959 | 2.526 | .4 | .6 | 0.4478 | 0.8942 | 0.5008 | 1.997 | .4 |
| .7 | 0.3697 | 0.9291 | 0.3979 | 2.513 | .3 | .7 | 0.4493 | 0.8934 | 0.5029 | 1.988 | .3 |
| .8 | 0.3714 | 0.9285 | 0.4000 | 2.500 | .2 | .8 | 0.4509 | 0.8926 | 0.5051 | 1.980 | .2 |
| .9 | 0.3730 | 0.9278 | 0.4020 | 2.488 | .1 | .9 | 0.4524 | 0.8918 | 0.5073 | 1.971 | .1 |
| 22.0 | 0.3746 | 0.9272 | 0.4040 | 2.475 | 68.0 | 27.0 | 0.4540 | 0.8910 | 0.5095 | 1.963 | 63.0 |
| .1 | 0.3762 | 0.9265 | 0.4061 | 2.463 | .9 | .1 | 0.4555 | 0.8902 | 0.5117 | 1.954 | .9 |
| .2 | 0.3778 | 0.9259 | 0.4081 | 2.450 | .8 | .2 | 0.4571 | 0.8894 | 0.5139 | 1.946 | .8 |
| .3 | 0.3795 | 0.9252 | 0.4101 | 2.438 | .7 | .3 | 0.4586 | 0.8886 | 0.5161 | 1.937 | .7 |
| .4 | 0.3811 | 0.9245 | 0.4122 | 2.426 | .6 | .4 | 0.4602 | 0.8878 | 0.5184 | 1.929 | .6 |
| .5 | 0.3827 | 0.9239 | 0.4142 | 2.414 | .5 | .5 | 0.4617 | 0.8870 | 0.5206 | 1.921 | .5 |
| .6 | 0.3843 | 0.9232 | 0.4163 | 2.402 | .4 | .6 | 0.4633 | 0.8862 | 0.5228 | 1.913 | .4 |
| .7 | 0.3859 | 0.9225 | 0.4183 | 2.391 | .3 | .7 | 0.4648 | 0.8854 | 0.5250 | 1.905 | .3 |
| .8 | 0.3875 | 0.9219 | 0.4204 | 2.379 | .2 | .8 | 0.4664 | 0.8846 | 0.5272 | 1.897 | .2 |
| .9 | 0.3891 | 0.9212 | 0.4224 | 2.367 | .1 | .9 | 0.4679 | 0.8838 | 0.5295 | 1.889 | .1 |
| 23.0 | 0.3907 | 0.9205 | 0.4245 | 2.356 | 67.0 | 28.0 | 0.4695 | 0.8829 | 0.5317 | 1.881 | 62.0 |
| .1 | 0.3923 | 0.9198 | 0.4265 | 2.344 | .9 | .1 | 0.4710 | 0.8821 | 0.5340 | 1.873 | .9 |
| .2 | 0.3939 | 0.9191 | 0.4286 | 2.333 | .8 | .2 | 0.4726 | 0.8813 | 0.5362 | 1.865 | .8 |
| .3 | 0.3955 | 0.9184 | 0.4307 | 2.322 | .7 | .3 | 0.4741 | 0.8805 | 0.5384 | 1.857 | .7 |
| .4 | 0.3971 | 0.9178 | 0.4327 | 2.311 | .6 | .4 | 0.4756 | 0.8796 | 0.5407 | 1.849 | .6 |
| .5 | 0.3987 | 0.9171 | 0.4348 | 2.300 | .5 | .5 | 0.4772 | 0.8788 | 0.5430 | 1.842 | .5 |
| .6 | 0.4003 | 0.9164 | 0.4369 | 2.289 | .4 | .6 | 0.4787 | 0.8780 | 0.5452 | 1.834 | .4 |
| .7 | 0.4019 | 0.9157 | 0.4390 | 2.278 | .3 | .7 | 0.4802 | 0.8771 | 0.5475 | 1.827 | .3 |
| .8 | 0.4035 | 0.9150 | 0.4411 | 2.267 | .2 | .8 | 0.4818 | 0.8763 | 0.5498 | 1.819 | .2 |
| .9 | 0.4051 | 0.9143 | 0.4431 | 2.257 | .1 | .9 | 0.4833 | 0.8755 | 0.5520 | 1.811 | .1 |
| 24.0 | 0.4067 | 0.9135 | 0.4452 | 2.246 | 66.0 | 29.0 | 0.4848 | 0.8746 | 0.5543 | 1.804 | 61.0 |
| .1 | 0.4083 | 0.9128 | 0.4473 | 2.236 | .9 | .1 | 0.4863 | 0.8738 | 0.5566 | 1.797 | .9 |
| .2 | 0.4099 | 0.9121 | 0.4494 | 2.225 | .8 | .2 | 0.4879 | 0.8729 | 0.5589 | 1.789 | .8 |
| .3 | 0.4115 | 0.9114 | 0.4515 | 2.215 | .7 | .3 | 0.4894 | 0.8721 | 0.5612 | 1.782 | .7 |
| .4 | 0.4131 | 0.9107 | 0.4536 | 2.204 | .6 | .4 | 0.4909 | 0.8712 | 0.5635 | 1.775 | .6 |
| .5 | 0.4147 | 0.9100 | 0.4557 | 2.194 | .5 | .5 | 0.4924 | 0.8704 | 0.5658 | 1.767 | .5 |
| .6 | 0.4163 | 0.9092 | 0.4578 | 2.184 | .4 | .6 | 0.4939 | 0.8695 | 0.5681 | 1.760 | .4 |
| .7 | 0.4179 | 0.9085 | 0.4599 | 2.174 | .3 | .7 | 0.4955 | 0.8686 | 0.5704 | 1.753 | .3 |
| .8 | 0.4195 | 0.9078 | 0.4621 | 2.164 | .2 | .8 | 0.4970 | 0.8678 | 0.5727 | 1.746 | .2 |
| .9 | 0.4210 | 0.9070 | 0.4642 | 2.154 | .1 | .9 | 0.4985 | 0.8669 | 0.5750 | 1.739 | .1 |
| 25.0 | 0.4226 | 0.9063 | 0.4663 | 2.145 | 65.0 | 30.0 | 0.5000 | 0.8660 | 0.5774 | 1.732 | 60.0 |
| | Cos θ | Sin θ | Cot θ | Tan θ | Degrees | | Cos θ | Sin θ | Cot θ | Tan θ | Degrees |

## TABLE D2  Trigonometric Values, Angle θ in Degrees and Tenths

| Degrees | Sin θ | Cos θ | Tan θ | Cot θ | | Degrees | Sin θ | Cos θ | Tan θ | Cot θ | |
|---|---|---|---|---|---|---|---|---|---|---|---|---|
| 30.0 | 0.5000 | 0.8660 | 0.5774 | 1.732 | 60.0 | 35.0 | 0.5736 | 0.8192 | 0.7002 | 1.428 | 55.0 |
| .1 | 0.5015 | 0.8652 | 0.5797 | 1.725 | .9 | .1 | 0.5750 | 0.8181 | 0.7028 | 1.423 | .9 |
| .2 | 0.5030 | 0.8643 | 0.5820 | 1.718 | .8 | .2 | 0.5764 | 0.8171 | 0.7054 | 1.418 | .8 |
| .3 | 0.5045 | 0.8634 | 0.5844 | 1.711 | .7 | .3 | 0.5779 | 0.8161 | 0.7080 | 1.412 | .7 |
| .4 | 0.5060 | 0.8625 | 0.5867 | 1.704 | .6 | .4 | 0.5793 | 0.8151 | 0.7107 | 1.407 | .6 |
| .5 | 0.5075 | 0.8616 | 0.5890 | 1.698 | .5 | .5 | 0.5807 | 0.8141 | 0.7133 | 1.402 | .5 |
| .6 | 0.5090 | 0.8607 | 0.5914 | 1.691 | .4 | .6 | 0.5821 | 0.8131 | 0.7159 | 1.397 | .4 |
| .7 | 0.5105 | 0.8599 | 0.5938 | 1.684 | .3 | .7 | 0.5835 | 0.8121 | 0.7186 | 1.392 | .3 |
| .8 | 0.5120 | 0.8590 | 0.5961 | 1.678 | .2 | .8 | 0.5850 | 0.8111 | 0.7212 | 1.387 | .2 |
| .9 | 0.5135 | 0.8581 | 0.5985 | 1.671 | .1 | .9 | 0.5864 | 0.8100 | 0.7239 | 1.381 | .1 |
| 31.0 | 0.5150 | 0.8572 | 0.6009 | 1.664 | 59.0 | 36.0 | 0.5878 | 0.8090 | 0.7265 | 1.376 | 54.0 |
| .1 | 0.5165 | 0.8563 | 0.6032 | 1.658 | .9 | .1 | 0.5892 | 0.8080 | 0.7292 | 1.371 | .9 |
| .2 | 0.5180 | 0.8554 | 0.6056 | 1.651 | .8 | .2 | 0.5906 | 0.8070 | 0.7319 | 1.366 | .8 |
| .3 | 0.5195 | 0.8545 | 0.6080 | 1.645 | .7 | .3 | 0.5920 | 0.8059 | 0.7346 | 1.361 | .7 |
| .4 | 0.5210 | 0.8536 | 0.6104 | 1.638 | .6 | .4 | 0.5934 | 0.8049 | 0.7373 | 1.356 | .6 |
| .5 | 0.5225 | 0.8526 | 0.6128 | 1.632 | .5 | .5 | 0.5948 | 0.8039 | 0.7400 | 1.351 | .5 |
| .6 | 0.5240 | 0.8517 | 0.6152 | 1.625 | .4 | .6 | 0.5962 | 0.8028 | 0.7427 | 1.347 | .4 |
| .7 | 0.5255 | 0.8508 | 0.6176 | 1.619 | .3 | .7 | 0.5976 | 0.8018 | 0.7454 | 1.342 | .3 |
| .8 | 0.5270 | 0.8499 | 0.6200 | 1.613 | .2 | .8 | 0.5990 | 0.8007 | 0.7481 | 1.337 | .2 |
| .9 | 0.5284 | 0.8490 | 0.6224 | 1.607 | .1 | .9 | 0.6004 | 0.7997 | 0.7508 | 1.332 | .1 |
| 32.0 | 0.5299 | 0.8480 | 0.6249 | 1.600 | 58.0 | 37.0 | 0.6018 | 0.7986 | 0.7536 | 1.327 | 53.0 |
| .1 | 0.5314 | 0.8471 | 0.6273 | 1.594 | .9 | .1 | 0.6032 | 0.7976 | 0.7563 | 1.322 | .9 |
| .2 | 0.5329 | 0.8462 | 0.6297 | 1.588 | .8 | .2 | 0.6046 | 0.7965 | 0.7590 | 1.317 | .8 |
| .3 | 0.5344 | 0.8453 | 0.6322 | 1.582 | .7 | .3 | 0.6060 | 0.7955 | 0.7618 | 1.313 | .7 |
| .4 | 0.5358 | 0.8443 | 0.6346 | 1.576 | .6 | .4 | 0.6074 | 0.7944 | 0.7646 | 1.308 | .6 |
| .5 | 0.5373 | 0.8434 | 0.6371 | 1.570 | .5 | .5 | 0.6088 | 0.7934 | 0.7673 | 1.303 | .5 |
| .6 | 0.5388 | 0.8425 | 0.6395 | 1.564 | .4 | .6 | 0.6101 | 0.7923 | 0.7701 | 1.299 | .4 |
| .7 | 0.5402 | 0.8415 | 0.6420 | 1.558 | .3 | .7 | 0.6115 | 0.7912 | 0.7729 | 1.294 | .3 |
| .8 | 0.5417 | 0.8406 | 0.6445 | 1.552 | .2 | .8 | 0.6129 | 0.7902 | 0.7757 | 1.289 | .2 |
| .9 | 0.5432 | 0.8396 | 0.6469 | 1.546 | .1 | .9 | 0.6143 | 0.7891 | 0.7785 | 1.285 | .1 |
| 33.0 | 0.5446 | 0.8387 | 0.6494 | 1.540 | 57.0 | 38.0 | 0.6157 | 0.7880 | 0.7813 | 1.280 | 52.0 |
| .1 | 0.5461 | 0.8377 | 0.6519 | 1.534 | .9 | .1 | 0.6170 | 0.7869 | 0.7841 | 1.275 | .9 |
| .2 | 0.5476 | 0.8368 | 0.6544 | 1.528 | .8 | .2 | 0.6184 | 0.7859 | 0.7869 | 1.271 | .8 |
| .3 | 0.5490 | 0.8358 | 0.6569 | 1.522 | .7 | .3 | 0.6198 | 0.7848 | 0.7898 | 1.266 | .7 |
| .4 | 0.5505 | 0.8348 | 0.6594 | 1.517 | .6 | .4 | 0.6211 | 0.7837 | 0.7926 | 1.262 | .6 |
| .5 | 0.5519 | 0.8339 | 0.6619 | 1.511 | .5 | .5 | 0.6225 | 0.7826 | 0.7954 | 1.257 | .5 |
| .6 | 0.5534 | 0.8329 | 0.6644 | 1.505 | .4 | .6 | 0.6239 | 0.7815 | 0.7983 | 1.253 | .4 |
| .7 | 0.5548 | 0.8320 | 0.6669 | 1.499 | .3 | .7 | 0.6252 | 0.7804 | 0.8012 | 1.248 | .3 |
| .8 | 0.5563 | 0.8310 | 0.6694 | 1.494 | .2 | .8 | 0.6266 | 0.7793 | 0.8040 | 1.244 | .2 |
| .9 | 0.5577 | 0.8300 | 0.6720 | 1.488 | .1 | .9 | 0.6280 | 0.7782 | 0.8069 | 1.239 | .1 |
| 34.0 | 0.5592 | 0.8290 | 0.6745 | 1.483 | 56.0 | 39.0 | 0.6293 | 0.7771 | 0.8098 | 1.235 | 51.0 |
| .1 | 0.5606 | 0.8281 | 0.6771 | 1.477 | .9 | .1 | 0.6307 | 0.7760 | 0.8127 | 1.230 | .9 |
| .2 | 0.5621 | 0.8271 | 0.6796 | 1.471 | .8 | .2 | 0.6320 | 0.7749 | 0.8156 | 1.226 | .8 |
| .3 | 0.5635 | 0.8261 | 0.6822 | 1.466 | .7 | .3 | 0.6334 | 0.7738 | 0.8185 | 1.222 | .7 |
| .4 | 0.5650 | 0.8251 | 0.6847 | 1.460 | .6 | .4 | 0.6347 | 0.7727 | 0.8214 | 1.217 | .6 |
| .5 | 0.5664 | 0.8241 | 0.6873 | 1.455 | .5 | .5 | 0.6361 | 0.7716 | 0.8243 | 1.213 | .5 |
| .6 | 0.5678 | 0.8231 | 0.6899 | 1.450 | .4 | .6 | 0.6374 | 0.7705 | 0.8273 | 1.209 | .4 |
| .7 | 0.5693 | 0.8221 | 0.6924 | 1.444 | .3 | .7 | 0.6388 | 0.7694 | 0.8302 | 1.205 | .3 |
| .8 | 0.5707 | 0.8211 | 0.6950 | 1.439 | .2 | .8 | 0.6401 | 0.7683 | 0.8332 | 1.200 | .2 |
| .9 | 0.5721 | 0.8202 | 0.6976 | 1.433 | .1 | .9 | 0.6414 | 0.7672 | 0.8361 | 1.196 | .1 |
| 35.0 | 0.5736 | 0.8192 | 0.7002 | 1.428 | 55.0 | 40.0 | 0.6428 | 0.7660 | 0.8391 | 1.192 | 50.0 |
| | Cos θ | Sin θ | Cot θ | Tan θ | Degrees | | Cos θ | Sin θ | Cot θ | Tan θ | Degrees |

**TABLE D2    Trigonometric Values, Angle θ in Degrees and Tenths**

| Degrees | Sin θ | Cos θ | Tan θ | Cot θ | |
|---|---|---|---|---|---|
| 40.0 | 0.6428 | 0.7660 | 0.8391 | 1.192 | 50.0 |
| .1 | 0.6441 | 0.7649 | 0.8421 | 1.188 | .9 |
| .2 | 0.6455 | 0.7638 | 0.8451 | 1.183 | .8 |
| .3 | 0.6468 | 0.7627 | 0.8481 | 1.179 | .7 |
| .4 | 0.6481 | 0.7615 | 0.8511 | 1.175 | .6 |
| .5 | 0.6494 | 0.7604 | 0.8541 | 1.171 | .5 |
| .6 | 0.6508 | 0.7593 | 0.8571 | 1.167 | .4 |
| .7 | 0.6521 | 0.7581 | 0.8601 | 1.163 | .3 |
| .8 | 0.6534 | 0.7570 | 0.8632 | 1.159 | .2 |
| .9 | 0.6547 | 0.7559 | 0.8662 | 1.154 | .1 |
| 41.0 | 0.6561 | 0.7547 | 0.8693 | 1.150 | 49.0 |
| .1 | 0.6574 | 0.7536 | 0.8724 | 1.146 | .9 |
| .2 | 0.6587 | 0.7524 | 0.8754 | 1.142 | .8 |
| .3 | 0.6600 | 0.7513 | 0.8785 | 1.138 | .7 |
| .4 | 0.6613 | 0.7501 | 0.8816 | 1.134 | .6 |
| .5 | 0.6626 | 0.7490 | 0.8847 | 1.130 | .5 |
| .6 | 0.6639 | 0.7478 | 0.8878 | 1.126 | .4 |
| .7 | 0.6652 | 0.7466 | 0.8910 | 1.122 | .3 |
| .8 | 0.6665 | 0.7455 | 0.8941 | 1.118 | .2 |
| .9 | 0.6678 | 0.7443 | 0.8972 | 1.115 | .1 |
| 42.0 | 0.6691 | 0.7431 | 0.9004 | 1.111 | 48.0 |
| .1 | 0.6704 | 0.7420 | 0.9036 | 1.107 | .9 |
| .2 | 0.6717 | 0.7408 | 0.9067 | 1.103 | .8 |
| .3 | 0.6730 | 0.7396 | 0.9099 | 1.099 | .7 |
| .4 | 0.6743 | 0.7385 | 0.9131 | 1.095 | .6 |
| .5 | 0.6756 | 0.7373 | 0.9163 | 1.091 | .5 |
| .6 | 0.6769 | 0.7361 | 0.9195 | 1.087 | .4 |
| .7 | 0.6782 | 0.7349 | 0.9228 | 1.084 | .3 |
| .8 | 0.6794 | 0.7337 | 0.9260 | 1.080 | .2 |
| .9 | 0.6807 | 0.7325 | 0.9293 | 1.076 | .1 |
| 43.0 | 0.6820 | 0.7314 | 0.9325 | 1.072 | 47.0 |
| .1 | 0.6833 | 0.7302 | 0.9358 | 1.069 | .9 |
| .2 | 0.6845 | 0.7290 | 0.9391 | 1.065 | .8 |
| .3 | 0.6858 | 0.7278 | 0.9424 | 1.061 | .7 |
| .4 | 0.6871 | 0.7266 | 0.9457 | 1.057 | .6 |
| .5 | 0.6884 | 0.7254 | 0.9490 | 1.054 | .5 |
| .6 | 0.6896 | 0.7242 | 0.9523 | 1.050 | .4 |
| .7 | 0.6909 | 0.7230 | 0.9556 | 1.046 | .3 |
| .8 | 0.6921 | 0.7218 | 0.9590 | 1.043 | .2 |
| .9 | 0.6934 | 0.7206 | 0.9623 | 1.039 | .1 |
| 44.0 | 0.6947 | 0.7193 | 0.9657 | 1.036 | 46.0 |
| .1 | 0.6959 | 0.7181 | 0.9691 | 1.032 | .9 |
| .2 | 0.6972 | 0.7169 | 0.9725 | 1.028 | .8 |
| .3 | 0.6984 | 0.7157 | 0.9759 | 1.025 | .7 |
| .4 | 0.6997 | 0.7145 | 0.9793 | 1.021 | .6 |
| .5 | 0.7009 | 0.7133 | 0.9827 | 1.018 | .5 |
| .6 | 0.7022 | 0.7120 | 0.9861 | 1.014 | .4 |
| .7 | 0.7034 | 0.7108 | 0.9896 | 1.011 | .3 |
| .8 | 0.7046 | 0.7096 | 0.9930 | 1.007 | .2 |
| .9 | 0.7059 | 0.7083 | 0.9965 | 1.003 | .1 |
| 45.0 | 0.7071 | 0.7071 | 1.0000 | 1.000 | 45.0 |
| | Cos θ | Sin θ | Cot θ | Tan θ | Degrees |

TABLE D3  Formulas

## General Use Formulas

**Percentage Gain and Loss:**

$$\% \text{ increase} = \frac{\text{increase}}{\text{original value}} (100\%)$$

$$\% \text{ loss} = \frac{\text{loss}}{\text{original value}} (100\%)$$

**Pythagorean Theorem**

For sides $a$, $b$, and $c$ of a right triangle, with longest side $c$,

$$c^2 = a^2 + b^2$$

## Plane Figures

**Triangle:**  Area $= \frac{1}{2}bh$

**Square:**  Area $= b^2$

**Rectangle:**  Area $= bh$

**Parallelogram:**  Area $= bh$

**Rhombus:**  Area $= bh$

**Triangle:**  Use Hero's formula.

$$s = \frac{1}{2}(a + b + c)$$

$$\text{Area} = \sqrt{s(s - a)(s - b)(s - c)}$$

**Trapezoid:**  Area $= \frac{1}{2}(a + b)h$

**Circle:**

$$\text{Area} = \pi r^2 = \frac{\pi d^2}{4}$$

$$\text{Circumference} = \pi d = 2\pi r$$

$$\text{Length of arc} = \frac{(\text{sector angle})}{360}(\text{circumference})$$

**Ellipse:**

$$\text{Area} = \pi ab \qquad a \text{ and } b \text{ are } \frac{1}{2} \text{ of axes}$$

$$\text{Perimeter} = \pi\sqrt{2a^2 + 2b^2}$$

## Solid Figures

**Rectangular Solid:**  $V = lwh$

**Right Prism:**

$$\text{Volume} = (\text{cross-sectional area})(\text{length})$$
$$\text{LSA} = (\text{perimeter of cross section})(\text{length})$$

**Pyramids and Cones:**

$$\text{Volume} = \frac{(\text{area of base})(\text{altitude})}{3}$$

$$\text{LSA} = \frac{1}{2}(\text{perimeter of base})(\text{slant height})$$

**Frustums of Pyramids and Cones:**

$$\text{Volume} = \frac{a(B + T + \sqrt{BT})}{3}$$

where

$T$ = area of top
$B$ = area of bottom
$a$ = altitude

$$\text{LSA} = (\text{average perimeter})(\text{slant height})$$

**Sphere:**

$$\text{Volume} = \frac{4\pi r^3}{3}$$

$$\text{Surface area} = 4\pi r^2$$

## Trigonometry

$$\sin(A) = \frac{\text{opposite side to angle } A}{\text{hypotenuse}}$$

$$\cos(A) = \frac{\text{adjacent side to angle } A}{\text{hypotenuse}}$$

$$\tan(A) = \frac{\text{opposite side to angle } A}{\text{adjacent side to angle } A}$$

**Law of Sines:**

$$\frac{a}{\sin(A)} = \frac{b}{\sin(B)} = \frac{c}{\sin(C)}$$

**Law of Cosines:**

$$a^2 = b^2 + c^2 - 2bc \cos(A)$$
$$b^2 = a^2 + c^2 - 2ac \cos(B)$$
$$c^2 = a^2 + b^2 - 2ab \cos(C)$$

**Area of a Triangle:**

$$\text{Area} = \frac{1}{2}ab \sin(C)$$

$$= \frac{1}{2}ac \sin(B)$$

$$= \frac{1}{2}bc \sin(A)$$

# E

## Answers to Odd-Numbered Exercises and All Self-Assessment Tests

### CHAPTER 1   COMMON FRACTIONS

**Self-Assessment Test No. 1 (Page 3)**

(1) $\dfrac{3}{2}$    (2) $\dfrac{11}{4}$    (3) $\dfrac{36}{5}$    (4) $\dfrac{65}{8}$    (5) $\dfrac{37}{12}$

(6) $\dfrac{25}{16}$    (7) $\dfrac{7}{3}$    (8) $\dfrac{89}{9}$    (9) $6\dfrac{1}{2}$    (10) $3\dfrac{3}{4}$

(11) $5\dfrac{2}{3}$    (12) $1\dfrac{3}{7}$    (13) $2\dfrac{2}{11}$    (14) $11\dfrac{3}{7}$    (15) $1\dfrac{2}{3}$

(16) $9\dfrac{4}{5}$

**Practice Exercise No. 1 (Pages 4–5)**

|  | Proper Fraction | Improper Fraction | Mixed Number |  | Proper Fraction | Improper Fraction | Mixed Number |
|---|---|---|---|---|---|---|---|
| $\dfrac{7}{8}$ | ✓ |  |  | $\dfrac{17}{3}$ |  | ✓ |  |
| $1\dfrac{1}{5}$ |  |  | ✓ | $\dfrac{18}{37}$ | ✓ |  |  |
| $\dfrac{7}{5}$ |  | ✓ |  | $3\dfrac{9}{64}$ |  |  | ✓ |

| | Proper Fraction | Improper Fraction | Mixed Number | | Proper Fraction | Improper Fraction | Mixed Number |
|---|---|---|---|---|---|---|---|
| $3\frac{1}{2}$ | | | ✓ | $4\frac{1}{12}$ | | | ✓ |
| $\frac{22}{7}$ | | ✓ | | $\frac{17}{7}$ | | ✓ | |
| $\frac{1}{4}$ | ✓ | | | $\frac{114}{113}$ | | ✓ | |
| $\frac{9}{5}$ | | ✓ | | $\frac{114}{115}$ | ✓ | | |
| $\frac{16}{9}$ | | ✓ | | $51\frac{2}{3}$ | | | ✓ |
| $2\frac{4}{5}$ | | | ✓ | $\frac{9}{75}$ | ✓ | | |
| $\frac{11}{20}$ | ✓ | | | $\frac{75}{9}$ | | ✓ | |
| $\frac{3}{8}$ | ✓ | | | $3\frac{11}{15}$ | | | ✓ |
| $7\frac{9}{11}$ | | | ✓ | $\frac{14}{15}$ | ✓ | | |
| $\frac{8}{5}$ | | ✓ | | $\frac{173}{100}$ | | ✓ | |
| $\frac{5}{32}$ | ✓ | | | $\frac{16}{19}$ | ✓ | | |
| $3\frac{9}{11}$ | | | ✓ | $2\frac{3}{8}$ | | | ✓ |

## Practice Exercise No. 2 (Page 4)

(1) $\frac{5}{2}$    (3) $\frac{6}{5}$    (5) $\frac{14}{5}$    (7) $\frac{13}{10}$    (9) $\frac{16}{9}$

## Practice Exercise No. 3 (Page 6)

(1) $\frac{9}{5}$    (3) $\frac{28}{3}$    (5) $\frac{11}{3}$    (7) $\frac{53}{6}$    (9) $\frac{10}{3}$

**(11)** $\dfrac{28}{15}$  **(13)** $\dfrac{8}{3}$  **(15)** $\dfrac{93}{8}$  **(17)** $\dfrac{101}{10}$  **(19)** $\dfrac{67}{12}$

## Practice Exercise No. 4 (Page 7)

**(1)** $1\dfrac{1}{2}$  **(3)** $2\dfrac{1}{4}$  **(5)** $1\dfrac{1}{5}$  **(7)** $4\dfrac{3}{4}$  **(9)** $6\dfrac{3}{10}$

**(11)** $6\dfrac{7}{8}$  **(13)** $4\dfrac{1}{4}$  **(15)** $1\dfrac{1}{7}$  **(17)** $2\dfrac{4}{5}$  **(19)** $5$

## Self-Assessment Test No. 2 (Page 8)

**(1)** $\dfrac{6}{8}$  **(2)** $\dfrac{15}{18}$ .  **(3)** $\dfrac{4}{10} = \dfrac{10}{25}$  **(4)** $\dfrac{49}{56}$

**(5)** $\dfrac{24}{30}$  **(6)** $\dfrac{42}{64}$  **(7)** $\dfrac{4}{5}$  **(8)** $\dfrac{3}{4}$

**(9)** $\dfrac{1}{2}$  **(10)** $\dfrac{4}{11}$  **(11)** $\dfrac{1}{4}$  **(12)** $\dfrac{7}{9}$

## Practice Exercise No. 5 (Page 9)

**(1)** $\dfrac{2}{4} = \dfrac{4}{8}$  **(3)** $\dfrac{6}{8} = \dfrac{15}{20}$  **(5)** $\dfrac{20}{25} = \dfrac{40}{50}$

**(7)** $\dfrac{24}{28} = \dfrac{30}{35}$  **(9)** $\dfrac{3}{9} = \dfrac{27}{81}$

## Practice Exercise No. 6 (Page 10)

**(1)** $\dfrac{8}{12} = \dfrac{2}{3}$  **(3)** $\dfrac{1}{4}$  **(5)** $\dfrac{2}{8} = \dfrac{1}{4}$  **(7)** $\dfrac{1}{3}$  **(9)** $\dfrac{4}{5}$

**(11)** $\dfrac{1}{5}$  **(13)** $\dfrac{1}{4}$  **(15)** $\dfrac{1}{3}$  **(17)** $\dfrac{43}{72}$  **(19)** $\dfrac{7}{12}$

**(21)** $\dfrac{11}{13}$  **(23)** $\dfrac{1}{2}$

## Self-Assessment Test No. 3 (Page 10)

**(1)** $\dfrac{11}{24}$  **(2)** $\dfrac{17}{32}$  **(3)** $4\dfrac{9}{20}$  **(4)** $2\dfrac{13}{15}$  **(5)** $1\dfrac{3}{16}$

**(6)** $4\dfrac{5}{12}$  **(7)** $5\dfrac{19}{20}$  **(8)** $4\dfrac{25}{32}$

## Practice Exercise No. 7 (Page 12)

(1) $\frac{3}{4}$     (3) $1\frac{4}{15}$     (5) $1\frac{3}{4}$     (7) $\frac{11}{14}$     (9) $1\frac{11}{30}$

(11) $\frac{1}{12}$     (13) $\frac{1}{5}$     (15) $\frac{11}{27}$     (17) $6\frac{2}{9}$     (19) $1\frac{1}{10}$

(21) $4\frac{1}{5}$     (23) $5\frac{3}{10}$     (25) $2\frac{9}{22}$     (27) $3\frac{11}{32}$     (29) $10\frac{5}{8}$

## Self-Assessment Test No. 4 (Page 13)

(1) $\frac{1}{12}$     (2) $\frac{7}{16}$     (3) $\frac{2}{5}$     (4) $\frac{1}{16}$     (5) 2

(6) $8\frac{3}{4}$     (7) 1     (8) 2     (9) 1     (10) $2\frac{1}{12}$

(11) 2     (12) $\frac{13}{32}$     (13) 5     (14) $1\frac{1}{4}$

## Practice Exercise No. 8 (Page 14)

(1) $\frac{1}{8}$     (3) $\frac{4}{45}$     (5) $\frac{5}{6}$     (7) $\frac{3}{4}$     (9) $\frac{4}{11}$

(11) $\frac{2}{11}$     (13) $\frac{3}{32}$     (15) $\frac{7}{16}$

## Practice Exercise No. 9 (Page 15)

(1) $10\frac{1}{2}$     (3) 180 mm     (5) 36     (7) \$400     (9) $115\frac{1}{2}$ lb

(11) \$3.50

## Practice Exercise No. 10 (Page 15)

(1) $2\frac{5}{8}$     (3) 1     (5) $1\frac{2}{7}$     (7) $1\frac{1}{5}$     (9) $\frac{15}{32}$

(11) $\frac{4}{7}$     (13) 750 m     (15) 520 ft

## Practice Exercise No. 11 (Page 16)

(1) $\frac{1}{2}$     (3) $1\frac{1}{14}$     (5) $\frac{3}{4}$     (7) $1\frac{2}{3}$     (9) $\frac{7}{16}$

(11) $\frac{35}{48}$     (13) $1\frac{1}{4}$     (15) $1\frac{1}{3}$

**Practice Exercise No. 12 (Page 17)**

(1) $\dfrac{1}{12}$    (3) $\dfrac{1}{5}$    (5) $\dfrac{1}{18}$    (7) $\dfrac{1}{15}$    (9) $\dfrac{1}{75}$

(11) $\dfrac{9}{14}$    (13) $\dfrac{1}{2}$    (15) 2    (17) $6\dfrac{1}{2}$    (19) $1\dfrac{3}{16}$

(21) $8\dfrac{1}{6}$    (23) $\dfrac{1}{2}$    (25) 12

**Practical Applications Exercise (Pages 17–25)**

(1)

| Size | A | B | C | D | E | F | G | H | J | K | L | M |
|------|---|---|---|---|---|---|---|---|---|---|---|---|
| S1 | | | $3\frac{1}{2}$ | | | $\frac{1}{4}$ | | $\frac{7}{8}$ | $1\frac{3}{4}$ | | | $\frac{1}{4}$ |
| S2 | | $2\frac{3}{8}$ | | $4\frac{5}{16}$ | $1\frac{15}{16}$ | | $2\frac{7}{8}$ | | | | | $\frac{19}{32}$ |
| S3 | $8\frac{15}{16}$ | | | $3\frac{13}{16}$ | | $1\frac{11}{16}$ | | | $1\frac{11}{16}$ | | 2 | |
| S4 | | $1\frac{1}{8}$ | | $3\frac{1}{16}$ | | $\frac{15}{16}$ | $1\frac{11}{16}$ | | | $3\frac{1}{4}$ | | |
| S5 | $6\frac{31}{32}$ | | | | $\frac{25}{32}$ | | $1\frac{1}{32}$ | | $1\frac{23}{32}$ | | $1\frac{1}{4}$ | |
| S6 | | | $6\frac{1}{16}$ | | | $1\frac{21}{32}$ | | $1\frac{9}{32}$ | $1\frac{27}{32}$ | | | $\frac{13}{32}$ |
| S7 | | $2\frac{1}{16}$ | | $6\frac{1}{16}$ | $1\frac{1}{8}$ | | $1\frac{9}{16}$ | | | | | $\frac{13}{32}$ |
| S8 | | | $\frac{9}{16}$ | | $\frac{21}{64}$ | | $\frac{9}{32}$ | | $\frac{5}{32}$ | | | $\frac{3}{16}$ |

(3)  (a) $13\frac{3}{4}''$    (b) $11''$    (c) $\frac{3}{8}''$    (d) 13

(5) $\frac{19''}{32}$, $1\frac{3}{4}''$

(7) $\frac{1''}{2}$, $1\frac{37''}{40}$        (9) $\frac{35''}{64}$        (11) 24

(13) \$43.75        (15) $\frac{1}{20}$, $\frac{1}{4}$

**(17)** **(a)** $\frac{30}{31}\,\Omega$  **(b)** $1\frac{5}{7}\,\Omega$  **(c)** $\frac{10}{13}\,\Omega$  **(d)** $5\frac{1}{3}\,\Omega$  **(e)** $18\,\Omega$  **(f)** $\frac{3}{5}\,\Omega$

## Miscellaneous Exercises (Pages 25–26)

**(1)** $6\frac{3}{16}$   **(3)** $\frac{5}{16}$   **(5)** $3\frac{5}{16}$   **(7)** $1\frac{31}{35}$   **(9)** $\frac{5}{24}$

**(11)** $\frac{2}{3}$   **(13)** $4''$

# CHAPTER 2   DECIMAL FRACTIONS

## Self-Assessment Test No. 1 (Page 27)

**(1)** 3.1   **(2)** 5.25   **(3)** 16.02   **(4)** 7.007   **(5)** 17.33

**(6)** 8.022   **(7)** $1\frac{3}{10}$   **(8)** $1\frac{4}{5}$   **(9)** $3\frac{7}{8}$   **(10)** $10\frac{3}{4}$

**(11)** $8\frac{9}{20}$   **(12)** $11\frac{3}{25}$

## Practice Exercise No. 1 (Page 28)

**(1)** 1.1   **(3)** 5.51   **(5)** 3.911   **(7)** 3.37   **(9)** 3.007
**(11)** 3.085   **(13)** 2.33   **(15)** 12.19

## Practice Exercise No. 2 (Page 29)

**(1)** $1\frac{1}{10}$   **(3)** $1\frac{111}{1000}$   **(5)** $1\frac{3}{1000}$   **(7)** $1\frac{3}{10}$   **(9)** $4\frac{2}{5}$
**(11)** $1\frac{93}{100}$   **(13)** $6\frac{1}{2}$   **(15)** $1\frac{3}{100}$   **(17)** $1\frac{937}{1000}$   **(19)** $16\frac{3}{4}$

## Practice Exercise No. 3 (Page 30)

**(1)** 7   **(3)** 700   **(5)** 413.4   **(7)** 612   **(9)** 476.3
**(11)** 4993   **(13)** 0.3   **(15)** 17.75   **(17)** 1230   **(19)** 666.7

## Practice Exercise No. 4 (Page 31)

**(1)** 0.9   **(3)** 0.442   **(5)** 14.0007   **(7)** 0.09   **(9)** 3.5
**(11)** 0.0475   **(13)** 11.001   **(15)** 0.0011   **(17)** 0.4421   **(19)** 4.075

## Practice Exercise No. 5 (Page 31)

**(1)** $\dfrac{17}{20}$  **(3)** $\dfrac{1}{2}$  **(5)** $\dfrac{99}{100}$  **(7)** $\dfrac{7}{8}$  **(9)** $\dfrac{9}{10}$

**(11)** $35\dfrac{23}{50}$  **(13)** $\dfrac{3}{25}$  **(15)** $\dfrac{1}{32}$  **(17)** $10\dfrac{9}{20}$  **(19)** $4\dfrac{39}{100}$

## Self-Assessment Test No. 2 (Page 32)

**(1)** 14.026  **(2)** 1004.135  **(3)** 17.085  **(4)** 272.13  **(5)** 0.99

**(6)** 100.43  **(7)** 0.988  **(8)** 12.863  **(9)** 0.67  **(10)** 11.925

## Practice Exercise No. 6 (Page 33)

**(1)** 27.186  **(3)** 492.11  **(5)** 8584.324  **(7)** 126.324

**(9)** 277.6276  **(11)** 190.271  **(13)** 148.96  **(15)** 181.223

**(17)** 260.489  **(19)** 1044.555

## Practice Exercise No. 7 (Pages 33, 34)

**(1)** 72.067  **(3)** 5.513  **(5)** 24.596  **(7)** 41.04

**(9)** 14.801  **(11)** 510.35  **(13)** 97.696  **(15)** 39.424

**(17)** 49.003  **(19)** 2.168

## Practical Applications
## Exercise No. 1 (Pages 34–36)

**(1)** 1.3°

**(3)** **(a)** 18.2 in.  **(b)** 9.1 in.  **(c)** 9.6 in.

**(5)** **(a)** 42.67 mm, 151.66 mm  **(b)** 17.09 mm

**(7)** **(a)** 0.2969 in., 0.2125 in., 0.8847 in., 0.5647 in., 0.3012 in., 0.3218 in.

**(b)** 0.1925 in., 0.1275 in.

**(c)** 5.5978 in.

## Self-Assessment Test No. 3 (Page 36)

**(1)** 6.1858  **(2)** 19.9305  **(3)** 257.1872  **(4)** 10.5315

**(5)** 21.125  **(6)** 1.456  **(7)** $9.35  **(8)** $22.51

**(9)** 520.7 mm  **(10)** 492.25 mm

## Practice Exercise No. 8 (Page 38)

**(1)** 0.777  **(3)** 80.99  **(5)** 1266.51  **(7)** 633.696

**(9)** 0.067 972  **(11)** 14.942 85  **(13)** 0.000 454 95  **(15)** 0.040 848

## Practical Applications
## Exercise No. 2 (Pages 38–39)

(1)

| Area (m²) | Cost ($) |
|-----------|----------|
| 0.99      | 2.71     |
| 2.4742    | 6.78     |
| 4.851     | 13.29    |
| 6.0444    | 16.56    |
| 8.5068    | 23.31    |

(3)   26 mm = 1.024″      30 mm = 1.182″
    49 mm = 1.931″      23 mm = 0.906″
    85 mm = 3.349″      98 mm = 3.861″
    57 mm = 2.246″

## Self-Assessment Test No. 4 (Page 39)

(1)   23.343      (2)   271      (3)   12.513      (4)   6.85      (5)   10.217
(6)   19.495

## Practice Exercise No. 9 (Page 41)

 (1)   459      (3)   2341      (5)   198      (7)   780      (9)   798
(11)   103

## Practice Exercise No. 10 (Page 42)

 (1)   0.9      (3)   4.32      (5)   0.673      (7)   1.67      (9)   3.69
(11)   2.56

## Practice Exercise No. 11 (Page 43)

 (1)   3.65      (3)   9.2      (5)   4.56      (7)   54.2      (9)   0.236
(11)   94

## Practice Exercise No. 12 (Page 45)

(1)   (a)   31.997      (b)   32.0      (c)   32.00
(3)   (a)   0.0373      (b)   0.04      (c)   0.037

## Practice Exercise No. 13 (Page 46)

 (1)   12.588      (3)   14.546      (5)   229.286      (7)   0.564
 (9)   179.570      (11)   94      (13)   0.048      (15)   10.850
(17)   1.451      (19)   2.627

**Self-Assessment Test No. 5 (Page 46)**

**(1)** 0.200 **(2)** 0.875 **(3)** 0.275
**(4)** 0.313 **(5)** 0.429 **(6)** 0.529
**(7)** $8\frac{12}{16}$ or $8\frac{3}{4}$ **(8)** $9\frac{14}{16}$ or $9\frac{7}{8}$
**(9)** $17\frac{34}{64}$ or $17\frac{17}{32}$ **(10)** $24\frac{51}{64}$

**Practice Exercise No. 14 (Page 47)**

**(1)** 0.2 **(3)** 0.875 **(5)** 0.125 **(7)** 0.8 **(9)** 0.6
**(11)** 0.3 **(13)** 0.093 **(15)** 0.79

**Practice Exercise No. 15 (Page 48)**

**(1)** 0.167 **(3)** 0.833 **(5)** 0.444 **(7)** 0.063 **(9)** 0.917
**(11)** 0.438

**Practice Exercise No. 16 (Page 49)**

**(1)** $12\frac{9}{16}$ **(3)** $3\frac{13}{16}$ **(5)** $12\frac{33}{64}$

**Practical Applications
Exercise No. 3 (Pages 49–52)**

**(1)** 0.625″, 0.0075″ **(3)** **(a)** 3.14 **(b)** 2.051
**(5)** $\frac{15″}{16}$, $\frac{63″}{64}$ **(7)** 0.024″ **(9)** 83.81 m
**(11)** **(a)** 13.396″ **(b)** 23.246″ **(c)** 39.4″ **(d)** 10.008″
**(e)** 315.2″ **(f)** 0.512″ **(g)** 0.296″ **(h)** 0.315″
**(13)** $152.19

**Practical Applications
Exercise No. 4 (Pages 52–53)**

**(1)** 10.2 kg **(3)** 264 **(5)** 24° **(7)** 29

**Miscellaneous Exercises (Pages 53–55)**

**(1)** 29.1 mm **(3)** 9.6″ **(5)** 36.3 **(7)** 0.875″
**(9)** 2.4 **(11)** 42.8 mm **(13)** 114.3 **(15)** **(a)** 87.71 **(b)** 83.05
**(17)** 68 lb **(19)** $0.72028 per kg
**(21)** 8.50 **(23)** 3

# CHAPTER 3 APPROXIMATION AND ESTIMATION

## Self-Assessment Test (Page 56)

**(1)** 16.8     **(2)** 191.3     **(3)** 6.75     **(4)** 0.0031     **(5)** 200 000

**(6)** 4000     **(7)** 20     **(8)** 2000     **(9)** 50     **(10)** 0.1

## Practice Exercise No. 1 (Pages 56–57)

**(1)** **(a)** 5.239 77     **(b)** 5.240     **(c)** 5.24     **(d)** 5.2

**(3)** **(a)** 0.005 00     **(b)** 0.0050     **(c)** 0.005     **(d)** 0.0

**(5)** **(a)** 110.9733     **(b)** 110.973     **(c)** 111.0     **(d)** 110.97

## Practice Exercise No. 2 (Pages 57–58)

**(1)** **(a)** 371.816     **(b)** 371.8     **(c)** 370     **(d)** 400

**(3)** **(a)** 44.714     **(b)** 44.7137     **(c)** 44.7     **(d)** 45

**(5)** **(a)** 0.006     **(b)** 0.006 17     **(c)** 0.006 174     **(d)** 0.006 173 9

## Practice Exercise No. 3 (Page 59)

**(1)** 90     **(3)** 2000     **(5)** 900     **(7)** 0.6

**(9)** 100     **(11)** 1     **(13)** 20 000 000     **(15)** 40

**(17)** 80     **(19)** 1000

## Miscellaneous Exercises (Page 60)

**(1)** **(a)** 21.79     **(b)** 22

**(3)** **(a)** 1.06 mm     **(b)** 1.06 mm

**(5)** **(a)** 85 miles     **(b)** 85.4 miles

# CHAPTER 4 PERCENTS

## Self-Assessment Test No. 1 (Page 61)

**(1)** 25%     **(2)** 20%     **(3)** 12.5%     **(4)** 25%

**(5)** 60%     **(6)** 70%     **(7)** 53%     **(8)** 66.7%

**(9)** 12%     **(10)** 34%     **(11)** $\dfrac{9}{20}$     **(12)** $\dfrac{77}{100}$

**(13)** $\dfrac{19}{20}$     **(14)** $\dfrac{7}{25}$     **(15)** $\dfrac{17}{200}$     **(16)** 0.39

**(17)**   0.64          **(18)**   0.95          **(19)**   0.75          **(20)**   0.2

## Practice Exercise No. 1 (Page 62)

**(1)**   50%          **(3)**   20%          **(5)**   62.5%          **(7)**   $33\frac{1}{3}\%$

**(9)**   10%          **(11)**   $55\frac{5}{9}\%$          **(13)**   16%          **(15)**   0.5%

**(17)**   $42\frac{6}{7}\%$

## Practice Exercise No. 2 (Page 62)

**(1)**   60%          **(3)**   87.5%          **(5)**   7.5%          **(7)**   95.5%

**(9)**   62.5%          **(11)**   0.2%          **(13)**   6.21%          **(15)**   71.9%

**(17)**   15.7%

## Practice Exercise No. 3 (Page 63)

**(1)**   $\frac{3}{10}$          **(3)**   $\frac{22}{25}$          **(5)**   $\frac{1}{8}$          **(7)**   $\frac{39}{100}$

**(9)**   $\frac{12}{25}$          **(11)**   $\frac{2}{3}$          **(13)**   $\frac{13}{100}$          **(15)**   $\frac{19}{20}$

**(17)**   $\frac{7}{8}$

## Practice Exercise No. 4 (Page 64)

**(1)**   0.37          **(3)**   0.185          **(5)**   0.0525          **(7)**   0.131 25

**(9)**   0.125          **(11)**   0.445          **(13)**   0.033 75          **(15)**   0.555

## Self-Assessment Test No. 2 (Page 64)

**(1)**   $30          **(2)**   $128          **(3)**   $3.96          **(4)**   88 mm

**(5)**   35.2 lb          **(6)**   50%          **(7)**   25%          **(8)**   $12\frac{1}{2}\%$

**(9)**   $2\frac{1}{2}\%$          **(10)**   0.625%

## Practice Exercise No. 5 (Page 66)

**(1)**   $12.60          **(3)**   $54          **(5)**   19.5 pt          **(7)**   0.38 km

**(9)**   0.3392 m          **(11)**   $600

## Practice Exercise No. 6 (Page 66)

(1)  $37\frac{1}{2}\%$

(3)  50%

(5)  1250%

(7)  $40\frac{5}{8}\%$

(9)  10%

## Practice Exercise No. 7 (Page 66)

(1)  320

(3)  45

## Practical Applications
## Exercise No. 1 (Pages 66–68)

(1)  26 min          (3)  3.825 m          (5)  $29 400

(7)

| Week | (a) Commission | (b) Wages |
|------|----------------|-----------|
| 1 | $375 | $825 |
| 2 | $235 | $685 |
| 3 | $597 | $1047 |
| 4 | $1400 | $1850 |
| 5 | $57.50 | $507.50 |

(9)  66

(11)  70%, 14%, 16%

(13)

| (a) discount | (b) cost |
|--------------|----------|
| 0 | $333 |
| $118.50 | $3831.50 |
| $1250 | $11 250 |
| $387.50 | $7362.50 |

## Self-Assessment Test No. 3 (Page 68)

(1)  50%          (2)  25%          (3)  $12\frac{1}{2}\%$          (4)  0.25%

(5)  15%

## Practical Applications
## Exercise No. 2 (Page 69)

(1)  20%          (3)  $4\frac{4}{9}\%$          (5)  20%          (7)  $16\frac{2}{3}\%$

## Miscellaneous Exercises (Page 70)

(**1**)  140 lb            (**3**)  $1.62            (**5**)  $99

(**7**)  (**a**) $6000    (**b**) $5100    (**c**) $4335            (**9**)  4%

(**11**)  4%

# CHAPTER 5   FUNDAMENTALS OF ALGEBRA

## Self-Assessment Test No. 1 (Page 73)

(**1**)  $a^5$            (**2**)  $b^{15}$            (**3**)  $a^5b^5$            (**4**)  $p^6q^4r^3$            (**5**)  $a^3$

(**6**)  $\dfrac{a^2}{b}$            (**7**)  $x\,y$            (**8**)  $p^3qr^2$            (**9**)  $x^6$            (**10**)  $p^8q^8$

(**11**)  $\dfrac{1}{81}x^8y^4$            (**12**)  4            (**13**)  2            (**14**)  3

(**15**)  1            (**16**)  1

## Practice Exercise No. 1 (Page 74)

(**1**)  $b^6$            (**3**)  $z^{12}$            (**5**)  $a^6b^3c^4$            (**7**)  $4^7$            (**9**)  $2^7$

(**11**)  $10^9$            (**13**)  $M^3N^5O^2$            (**15**)  $6\,A^3B^7$

## Practice Exercise No. 2 (Page 75)

(**1**)  $z$            (**3**)  $b^4$            (**5**)  $a^2b^4c^2$            (**7**)  $x^3y^3$            (**9**)  $\dfrac{1}{3}\dfrac{z\,x^{13}}{y^4}$

## Practice Exercise No. 3 (Page 76)

(**1**)  $x^8$            (**3**)  $z^{20}$            (**5**)  $x^{18}y^{18}z^2$            (**7**)  $x^{30}z^{20}$            (**9**)  $\dfrac{16}{9}a^6b^8$

(**11**)  $\dfrac{x^8y^8}{a^4b^6}$

## Practice Exercise No. 4 (Page 77)

(**1**)  $\dfrac{1}{z^4}$            (**3**)  $\dfrac{a^2}{b^4}$            (**5**)  $\dfrac{y^4z^5}{x^3}$            (**7**)  $b^4$            (**9**)  $\dfrac{3p^3q^4}{2MN}$

## Practice Exercise No. 5 (Page 77)

(**1**)  7            (**3**)  3            (**5**)  1            (**7**)  12            (**9**)  10

(**11**)  4

## Practice Exercise No. 6 (Page 78)

**(1)** $a^9b^8$     **(3)** $a^2b^4c^2$     **(5)** $b^2$     **(7)** $a^{24}$     **(9)** $x^3y^4z^{-3}$

## Self-Assessment Test No. 2 (Page 78)

**(1)** $+5$     **(2)** $-1$     **(3)** $+1$     **(4)** $+3$     **(5)** $-2$

**(6)** $+8$     **(7)** $-1$     **(8)** $+6$     **(9)** $-12$     **(10)** $-3$

**(11)** $+3$     **(12)** $+2$

## Practice Exercise No. 7 (Page 79)

**(1)** $+5$     **(3)** $+7$     **(5)** $-1$     **(7)** $+5$     **(9)** $+6$

**(11)** $+1$     **(13)** $-4$     **(15)** $-2$     **(17)** $+8$     **(19)** $+2$

## Practice Exercise No. 8 (Page 80)

**(1)** $+6$     **(3)** $-12$     **(5)** $-3$     **(7)** $+30$     **(9)** $+9$

**(11)** $+16$     **(13)** $-3$     **(15)** $+1\frac{1}{2}$     **(17)** $-4$     **(19)** $-3$

**(21)** $+\frac{1}{2}$     **(23)** $-4$

## Self-Assessment Test No. 3 (Page 81)

**(1)** $10a$     **(2)** $5x + 3y$     **(3)** $9z - 2q$     **(4)** $5p^2 + p$

**(5)** $7a^2 + 12a$     **(6)** $13ab + 6b$     **(7)** $18xy - 9y$     **(8)** $25x^3 - 15x^2$

## Practice Exercise No. 9 (Page 82)

**(1)** $10a$     **(3)** $8z + 2x - 3y$   **(5)** $-p + q$     **(7)** $17x^2 - 6x$

**(9)** $8y - y^2$     **(11)** $13ab - 4cd$    **(13)** $-2x - xy$     **(15)** $9c^2 + 12c$

## Self-Assessment Test No. 4 (Page 82)

**(1)** $x^2 - x$     **(2)** $ab + ac$     **(3)** $y^3 + y^2 - 4y$   **(4)** $x^2y - xy^2$

**(5)** $xyz + x^2yz$     **(6)** $ab^2 - abc$     **(7)** $pqr - pqs$     **(8)** $x^5y^2 + x^2y^4$

**(9)** $x^2 + x - 6$

## Practice Exercise No. 10 (Page 82)

**(1)** $3a + 6$     **(3)** $6a + 6b - 6c$     **(5)** $ab - 6a$

**(7)** $15x + 15y - 15z$     **(9)** $xyz - 2xy$     **(11)** $17x^2 - 17x + 51$

## Miscellaneous Exercises (Page 83)

**(1)** **(a)** 72         **(b)** 1000         **(c)** $6x^5$

**(3)** $\dfrac{1}{3}$

**(5)** $3x^2 + 3x - 8$

**(7)** $ax^2 - ax$

**(9)** $-x + 13$   or   $13 - x$

**(11)** $2x^2 + 3x + 1$

**(13)** $A^2 + 4A$

**(15)** $ab + b^2$

# CHAPTER 6    FORMULAS AND EQUATIONS

## Self-Assessment Test No. 1 (Page 84)

**(1)** 2         **(2)** 4         **(3)** 7         **(4)** 6         **(5)** $-2$         **(6)** 14

## Practice Exercise No. 1 (Page 85)

**(1)** 5              **(3)** 9              **(5)** 12              **(7)** 6

**(9)** 3              **(11)** 15              **(13)** 125              **(15)** $\dfrac{1}{2}$

## Practical Applications Exercise No. 1 (Pages 85–88)

**(1)** **(a)** 294.7 rev/min         **(b)** 627.6 rev/min         **(c)** 129.9 rev/min

**(3)** **(a)** 12.06         **(b)** 10.72         **(c)** 77.21         **(d)** 3.02

**(5)** 0.304″

## Practical Applications Exercise No. 2 (Pages 90–92)

**(1)** **(a)** $52x$     **(b)** $48x^2$     **(c)** 104 in., 192 in.²     **(d)** 1560 mm, 43 200 mm²     **(e)** 520 in., 4800 in.²     **(f)** 1346.8 mm, 32 199 mm²
**(g)** 139.36 in., 344.8 in.²

**(3)** **(a)** $2\pi r(h + r)$     **(b)** $\pi r^2 h$     **(c)** 534.14 cm², 942.6 cm³
**(d)** 273.4 in.², 325.2 in.³

## Self-Assessment Test No. 2 (Page 92)

**(1)** 4 **(2)** 7 **(3)** 3 **(4)** 33 **(5)** 36 **(6)** 7.5
**(7)** 22 **(8)** 30 **(9)** 0.48 **(10)** 9

## Practice Exercise No. 2 (Page 93)

**(1)** 16 **(3)** $-2$ **(5)** 20 **(7)** 5 **(9)** 5
**(11)** 1.3 **(13)** $\dfrac{8}{5}$ **(15)** 4 **(17)** 4

## Practice Exercise No. 3 (Page 96)

**(1)** 5 **(3)** 2 **(5)** 17 **(7)** $-6$ **(9)** $-\dfrac{1}{4}$

## Practical Applications Exercise No. 3 (Pages 96–98)

**(1)** 12 cm **(3) (a)** 8 mm **(b)** 70 in.
**(5) (a)** 200 **(b)** 50 **(c)** 132 **(d)** 63 **(e)** 78
**(7) (a)** 2.08 A **(b)** 24 V **(9)** 45 Ω
**(11) (a)** 5.64 cm **(b)** 1.30 yd

## Self-Assessment Test No. 3 (Page 98)

**(1)** $a = x - 4$ **(2)** $a = 5 - x$ **(3)** $v = \dfrac{500}{p}$ **(4)** $p = 5x$

**(5)** $c = \dfrac{a+3}{b}$ **(6)** $r = \dfrac{5}{2\pi}$ **(7)** $y = \dfrac{4}{x^2}$ **(8)** $x = \sqrt{\dfrac{4}{y}}$

**(9)** $x = 3t + y$ **(10)** $y = x - 3t$

## Practice Exercise No. 4 (Page 99)

**(1)** $c = a - b$ **(3)** $x = 4s$ **(5)** $s = \dfrac{4}{7t}$ **(7)** $y = \dfrac{4z}{x}$

**(9)** $p = zt - 7$

## Practical Applications Exercise No. 4 (Pages 99–101)

**(1) (a)** $S = \dfrac{\pi ND}{12}$ **(b)** $D = \dfrac{12S}{\pi N}$ **(c)** 763.9 **(d)** 152.8

**(3)** **(a)** $F = \dfrac{9C}{5} + 32$    **(b)** 208.4°    **(c)** 113°    **(d)** 32°

**(5)** **(a)** $x = \dfrac{100}{45}(W - 15)$    **(b)** $54.15    **(c)** 145    **(d)** 179

## Miscellaneous Exercises (Pages 101–103)

**(1)** 3

**(3)** $I = \sqrt{\dfrac{P}{R}}$, 4.5

**(5)** $v = \sqrt{\dfrac{pa}{ks}}$, $2\frac{1}{3}$

**(7)** $L = \dfrac{4V}{\pi D^2}$

**(9)** **(a)** $r = \sqrt{\dfrac{V}{\pi h}}$    **(b)** 34

**(11)** **(a)** 8    **(b)** 6
    **(c)** 4.5    **(d)** 0.75

**(13)** $g = \dfrac{W(n + x)}{mh}$

**(15)** 3

**(17)** $E = \dfrac{I(R + nr)}{n}$

**(19)** $H = \dfrac{A}{2\pi R} - \dfrac{R}{2}$

**(21)** 2

**(23)** $t(L - t + B)$

## CHAPTER 7    RATIO AND PROPORTION

### Self-Assessment Test (Page 104)

**(1)** 1:2    **(2)** 3:1    **(3)** 5:1    **(4)** 1:16

**(5)** 2:1    **(6)** $70, $210, $420    **(7)** $210, $420, $770

**(8)** 30°, 90°, 240°    **(9)** 63    **(10)** 112

### Practice Exercise No. 1 (Page 105)

**(1)** 1:10    **(3)** 1:20    **(5)** 6:7    **(7)** 3:1    **(9)** 4:5

**(11)** 7:3    **(13)** 21:2    **(15)** 1:12    **(17)** 2:1

### Practical Applications
### Exercise No. 1 (Page 107)

**(1)** $665, $285    **(3)** $8, $28, $52    **(5)** **(a)** 1.96 lb    **(b)** 375 lb

**(7)** **(a)** 8 qt, 15 qt    **(b)** 33.75 qt    **(9)** 0.3 m³, 0.9 m³

### Practical Applications
### Exercise No. 2 (Pages 109–110)

**(1)** $200    **(3)** 112 miles    **(5)** 0.011 25″    **(7)** $21.67

**(9)** 0.162″    **(11)** 5.415 Ω

## Practical Applications
### Exercise No. 3 (Pages 112–113)

**(1)**  4          **(3)**  35          **(5)**  240, 192          **(7)**  105          **(9)**  12

## Miscellaneous Exercises (Pages 113–114)

**(1)**  30 mm, 60 mm, 90 mm          **(3)**  147 oz, 63 oz      **(5)**  68 lb, 32 lb

**(7)**  $324, $288, $252, $180          **(9)**  3000 rev/min    **(11)**  80 yd³/s


## CHAPTER 8   SQUARES, SQUARE
ROOTS AND
THE THEOREM
OF PYTHAGORAS

## Self-Assessment Test No. 1 (Page 117)
(To 4 significant digits)

**(1)**  21.82          **(2)**  87.89          **(3)**  221.1          **(4)**  3133

**(5)**  1 221 000          **(6)**  0.001 88          **(7)**  0.000 548

**(8)**  0.000 065 61

## Practice Exercise No. 1 (Page 118)

**(1)**  479.6          **(3)**  282.2          **(5)**  4697          **(7)**  $1.742 \times 10^{-8}$

**(9)**  0.019 32          **(11)**  18 770          **(13)**  7921          **(15)**  10 820

**(17)**  1.082          **(19)**  0.005 402

## Self-Assessment Test No. 2 (Page 118)

**(1)**  6.914          **(2)**  8.328          **(3)**  100.9          **(4)**  128.5

**(5)**  0.9752          **(6)**  0.9355          **(7)**  0.032 09          **(8)**  0.2114

## Practice Exercise No. 2 (Page 119)

**(1)**  2.477          **(3)**  9.45          **(5)**  83.25          **(7)**  97.52

**(9)**  9.882          **(11)**  8.171          **(13)**  13.10          **(15)**  40.16

## Practice Exercise No. 3 (Page 120)

**(1)**  0.4359          **(3)**  0.4438          **(5)**  0.9149          **(7)**  0.9220

**(9)**  0.3633          **(11)**  0.3056          **(13)**  0.2958          **(15)**  0.060 83

## Self-Assessment Test No. 3 (Page 120)

(**1**)   16.34 cm  (**2**)  9.981″  (**3**)  120 mm  (**4**)  8.371 ft
(**5**)   0.435″

## Practice Exercise No. 4 (Page 123)

(**1**)  4.191 m  (**3**)  211.4 mm  (**5**)  10.92″  (**7**)  12.67 ft
(**9**)  4.409 yd

## Practical Applications Exercise (Pages 124–127)

(**1**)  (**a**)  5 yd  (**b**)  10.63 yd  (**3**)  2.45 ft  (**5**)  3.77 m
(**7**)  135.03 yd  (**9**)  74.18 mm  (**11**)  (**a**)  3.69 cm  (**b**)  3.29 cm

## Miscellaneous Exercises (Pages 127–131)

(**1**)  Side of square 141.4 mm  (**3**)  1 mm
(**5**)  (**a**)  240 cm  (**b**)  63.64 cm  (**7**)  59.8″

# CHAPTER 9   AREAS OF
             PLANE FIGURES

## Self-Assessment Test No. 1 (Page 132)

(**1**)  270 ft²  (**2**)  170 cm²  (**3**)  38 in.²  (**4**)  4.05 m²

## Practice Exercise No. 1 (Pages 135–136)

(**1**)  9  (**3**)  36  (**5**)  14

## Practice Exercise No. 2 (Pages 136–137)

(**1**)  36 yd²  (**3**)  81 in.²  (**5**)  22 yd²  (**7**)  15 in.²

## Practice Exercise No. 3 (Pages 138–140)

(**1**)  Total area = 554  (**3**)  Area of shape = 213  (**5**)  678 cm²

## Practice Exercise No. 4 (Page 141)

(**1**)  27.71 cm²  (**3**)  59.92 mm²  (**5**)  9501 cm² or 0.9501 m²

## Practical Applications
### Exercise No. 1 (Pages 142–145)

**(1)** 269.5 yd²        **(3)** 1800 in.², $5.40       **(5)** 2550 bricks

**(7)** 198 in.²        **(9)** **(a)** 3.76 m²     **(b)** 182.47 kg

**(11)** 12 297.2 lb

## Self-Assessment Test No. 2 (Page 146)

**(1)** 32 cm²     **(2)** 15 000 cm²     **(3)** 1728 in.²     **(4)** $1\frac{1}{12}$ ft²

**(5)** 1.07 yd²     **(6)** 113.26 ft²     **(7)** 1 m²     **(8)** 160 000 cm²

## Practice Exercise No. 5 (Page 148)

**(1)** **(a)** 50 000 cm²     **(b)** 70 000 cm²     **(c)** 43 000 cm²
      **(d)** 47 500 cm²     **(e)** 46 300 cm²

**(3)** **(a)** 397.44 in²     **(b)** 528.48 in²     **(c)** 1268.64 in²
      **(d)** 138.24 in²     **(e)** 171.36 in²

## Practical Applications
### Exercise No. 2 (Pages 148–149)

**(1)** **(a)** 94.5 yd²     **(b)** $897.75

**(3)** **(a)** 79.5 cm²     **(b)** 0.007 95 m²

## Miscellaneous Exercises (Pages 149–152)

**(1)** 352 mm²     **(3)** 11.25 yd²     **(5)** 0.25 m²     **(7)** 3.8 rolls (4 rolls)

**(9)** **(a)** 6     **(b)** 5 cm     **(c)** 120 000 **(11)** 3600 mm²     **(13)** 1200 mm²

## CHAPTER 10    THE CIRCLE
##                  AND THE ELLIPSE

## Self-Assessment Test No. 1 (Page 153)

**(1)** 25.14 cm     **(2)** 11.00 mm     **(3)** 59.70″     **(4)** 141.4 ft

**(5)** 85.9°     **(6)** 200.5°     **(7)** 30.55″     **(8)** 51.32 mm

## Practice Exercise No. 1 (Page 155)

**(1)** 27.02 cm     **(3)** 301.6 mm     **(5)** 197.9 mm

## Practical Applications Exercise No. 1 (Pages 155–158)

**(1)** **(a)** 659.8 cm **(b)** 6.598 m **(c)** 6598 mm **(3)** 628.48 cm
**(5)** 100.54 cm, estimate = 30 (true answer = 29.84) **(7)** 5499 in.

## Practice Exercise No. 2 (Pages 159–160)

**(1)** 15.71 mm **(3)** 82.91 mm **(5)** 70.70 cm

## Practice Exercise No. 3 (Pages 160–161)

**(1)** 80.2° **(3)** 200.5° **(5)** 329.4°

## Practical Applications Exercise No. 2 (Pages 161–163)

**(1)** 302.8 cm **(3)** 9.093 m **(5)** 62.84 yd **(7)** 28.53″

## Self-Assessment Test No. 2 (Page 164)

**(1)** 1809.8 cm² **(2)** 17 674 in.² **(3)** 283.56 m²
**(4)** 15.91 ft² **(5)** 17.45 cm² **(6)** 72.61 in.²
**(7)** 3211.8 mm² **(8)** 58.32 yd²

## Practice Exercise No. 4 (Page 166)

**(1)** 113.11 cm² **(3)** 5027.2 mm² **(5)** 3217.4 mm²
**(7)** 176.73 mm²

## Practice Exercise No. 5 (Page 167)

**(1)** 13.09 cm² **(3)** 4.61 in.² **(5)** 1027.8 mm²

## Practical Applications Exercise No. 3 (Pages 167–169)

**(1)** 628.4 in.² **(3)** 504.54 ft² **(5)** 12 137 yd²
**(7)** 856.4 cm²

## Practice Exercise No. 6 (Page 171)

| | Areas | Perimeters |
|---|---|---|
| **(1)** | 14.26 cm² | 34.70 cm |
| **(3)** | 94.26 mm² | 37.77 mm |

## Miscellaneous Exercises (Page 171–177)

(1)  4525 in.² (approx.)      (3)  769.8 mm²      (5)  $\pi\left(dh + \frac{1}{4}D^2\right)$

(7)  12.3 m²      (9)  212.7 mm      (11)  653 mm²      (13)  7.28″

(15)  (a)  12.6 ft      (b)  125.7 ft²      (17)  (a)  38.5 yd²      (b)  120

(19)  232.2 mm,  464.3 mm

# CHAPTER 11   SOLID FIGURES

## Self-Assessment Test (Page 178)

(1)  1050 mm³      (2)  15 cm³      (3)  84 in.³

(4)  62 840 in.³

## Practice Exercise No. 1 (Page 180)

(1)  108 cm³      (3)  35 mm³      (5)  298.89 cm³

## Practice Exercise No. 2 (Page 182)

(1)  9477      (3)  648 000      (5)  1.407      (7)  7.383

## Practice Exercise No. 3 (Page 184)

(1)  Yes      (3)  No      (5)  No

## Practice Exercise No. 4 (Pages 186–188)

(1)  22 000 mm³,  1100 mm²      (3)  44 280 cm³,  8778 cm²

(5)  270 888 mm³,  35 112 mm²      (7)  10 369 mm³

(9)  116 000 mm³      (11)  655.2 cm³

## Practice Exercise No. 5 (Pages 192–193)

|  | Volume | LSA |
|---|---|---|
| (1) | 96 in.³ | 102.6 in.² |
| (3) | 166.7 in.³ | 141.4 in.² |
|  | Volume |  |
| (5) | 133.33 in.³ |  |

## Practice Exercise No. 6 (Page 196)

**(1)** Volume = 158.1 cm³

**(3)** Volume = 238.3 mm³,   LSA = 173.0 mm²

## Practice Exercise No. 7 (Page 197)

|       | **Volume** | **Surface Area** |
|-------|------------|------------------|
| **(1)** | 113.1 in.³ | 113.1 in.² |
| **(3)** | 268.1 ft³ | 201.1 ft² |

## Practical Applications Exercise (Pages 197–200)

**(1)** **(a)** 5000 ft³   **(b)** 37 400 gal       **(3)** 419.69 cm³

**(5)** 48.38 yd           **(7)** 344.3 m³       **(9)** 9.6 yd

**(11)** **(a)** 42.42   **(b)** 42 420   **(c)** 29.82 m²

**(13)** **(a)** 3660.53 kg       **(b)** No

**(15)** 377 490 g   or   377.49 kg

**(17)** 4.596 m²  (slant height = 2.419)

## Miscellaneous Exercises (Pages 200–204)

**(1)** **(a)** 188 580   **(b)** 93.78%   **(3)** 6000 mm³   **(5)** 18 480 mm³

**(7)** 254.6 in.³ or 0.050 yd³       **(9)** 350   **(11)** 3 yd³

**(13)** **(a)** 1654 mm³   **(b)** 52.7%   **(15)** 3142 in.³

**(17)** **(a)** **(i)** 5600 cm²   **(ii)** 4400 cm²
    **(b)** **(i)** 19 600 cm³   **(ii)** 15 400 cm³
    **(c)** Same

# CHAPTER 12   GEOMETRICAL CONSTRUCTIONS

The answers for this chapter require constructions. Answers are not given unless specific values are requested. Where appropriate, references to Chapter 12 are made to assist the reader (e.g., ref. sect. 2 means refer to section 2 in Chapter 12 for method).

## Self-Assessment Test (Page 205)

**(1)** ref. sect. 6       **(2)** ref. sect. 10, 12, 13

**(3)** ref. sect. 9       **(4)** ref. sect. 14       **(5)** ref. sect. 6, 7, 14

## Practice Exercise (Page 213)

**(1)**   ref. sect. 1                    **(3)**   ref. sect. 9                    **(5)**   ref. sect. 10

**(7)**   ref. sect. 10, 12, 13          **(9)**   ref. sect 16

## Practical Applications Exercise (Pages 213–218)

**(1)**   ref. sect. 10, 13          **(3)**   ref. sect. 13;   $2\frac{21''}{32}$,   $6\frac{1''}{2}$

**(5)**   ref. sect. 14              **(7)**   ref. sect. 16                    **(9)**   ref. sect. 9

## Miscellaneous Exercises (Page 218)

**(1)**   ref. sect. 11              **(3)**   ref. sect. 6

# CHAPTER 13   ANGLES

## Self-Assessment Test No. 1 (Page 221)

**(1)**   29°51′23″              **(2)**   115°17′9″              **(3)**   94°55′52″              **(4)**   105°27′4″
**(5)**   11°40′15″              **(6)**   13°59′35″              **(7)**   2°31′59″              **(8)**   85°0′1″
**(9)**   87°30′36″              **(10)**   18.2542°

## Practice Exercise No. 1 (Page 222)

**(1)**   41°45′              **(3)**   169°13′              **(5)**   88°13′37″              **(7)**   130°56′14″
**(9)**   3°55′46″            **(11)**   15°9′              **(13)**   15°2′28″              **(15)**   6°42′16″

## Practice Exercise No. 2 (Page 224)

**(1)**   17°30′36″              **(3)**   47°18′              **(5)**   22.3°              **(7)**   88.1375°

## Self-Assessment Test No. 2 (Page 224)

**(4)**   $a = 110°$, $b = 50°$, $c = 20°$          **(5)**   $d = 40°$, $e = 50°$
**(6)**   $f = 40°$, $g = 40°$, $h = 50°$          **(7)**   $i = 30°$, $j = 60°$, $k = 50°$
**(8)**   $l = 40°$, $m = 70°$, $n = 60°$, $p = 50°$

## Practice Exercise No. 6 (Pages 233–234)

**(1)**   62°              **(3)**   70°              **(5)**   $g = 50°$, $h = 130°$, $i = 50°$
**(7)**   $m = 40°$, $n = 10°$, $p = 40°$          **(9)**   $u = 80°$, $v = 49°$, $w = 51°$

**Miscellaneous Exercises (Pages 234–236)**

**(1)** **(a)** 135°     **(b)** 195 000 mm²          **(3)** **(a)** 64°·     **(b)** *QR*          **(5)** 98°

# CHAPTER 14   TRIGONOMETRY OF RIGHT TRIANGLES

**Self-Assessment Test No. 1 (Page 237)**

**(1)** 32°24′          **(2)** 40°54′          **(3)** 33°52′          **(4)** 45°53′

**Practice Exercise No. 1 (Page 239)**

| Angle | Sine | Cosine | Tangent |
|-------|------|--------|---------|
| 50.7° | 0.7738 | 0.6333 | 1.2218 |
| 30° | 0.5000 | 0.8660 | 0.5774 |
| 60° | 0.8660 | 0.5000 | 1.7321 |
| 10.2° | 0.1771 | 0.9842 | 0.1799 |
| 18.9° | 0.3239 | 0.9461 | 0.3424 |
| 82.9° | 0.9923 | 0.1236 | 8.028 |
| 21.2° | 0.3616 | 0.9323 | 0.3879 |
| 33.7° | 0.5548 | 0.8320 | 0.6669 |

**Practice Exercise No. 2 (Page 240)**

| Angle | Sine | Cosine | Tangent |
|-------|------|--------|---------|
| 38.7° | 0.6252 | 0.7804 | 0.8012 |
| 64.7° | 0.9041 | 0.4274 | 2.116 |
| 71.3° | 0.9472 | 0.3206 | 2.954 |
| 41.8° | 0.6665 | 0.7455 | 0.8941 |
| 53.5° | 0.8039 | 0.5948 | 1.351 |
| 82.7° | 0.9919 | 0.1271 | 7.806 |
| 4° | 0.0698 | 0.9976 | 0.0699 |
| 87.4° | 0.9990 | 0.0454 · | 22.02 |
| 3.7° | 0.0645 | 0.9979 | 0.0647 |

**Practice Exercise No. 3 (Pages 242–243)**

**(1)** 19.5°          **(3)** 17.5°          **(5)** 43.1°          **(7)** 51.3°

## Practice Exercise No. 4 (Pages 243–244)

**(1)** 17.6°  **(3)** 73.5°  **(5)** 33.1°  **(7)** 42.1°

## Self-Assessment Test No. 2 (Page 244)

**(1)** 4.589 yd  **(2)** 10.392 m  **(3)** 46.36 mm  **(4)** 6.43″

## Practice Exercise No. 5 (Pages 246–247)

**(1)** $a = 2.108$ cm, $b = 3.400$ cm  **(3)** 3.68 mm
**(5)** $g = 71.33$ mm, $h = 70.09$ mm  **(7)** 1.04 mm

## Practice Exercise No. 6 (Page 248)

**(1)** 52.849 cm  **(3)** 30.28 mm  **(5)** 31.213″

## Practical Applications Exercise (Pages 249–254)

**(1)** 7.498″  **(3)** 35.7°
**(5)** $x_1 = y_2 = 1.945″$, $x_2 = 1.375″$, $y_1 = 2.382″$  **(7)** 0.249″
**(9)** 8.2°  **(11) (a)** 3.288 in.²  **(b)** 1.97″  **(13)** 10.7°

## Miscellaneous Exercises (Pages 254–264)

**(1) (a)** 60°  **(b)** 3.464 m  **(3)** 62.34 mm
**(5) (a)** 680 cm  **(b)** 80 cm, 28°  **(c)** 88 cm
**(7) (a)** 4′8″  **(b)** 8′8″  **(c)** 41.6°  **(d)** 86″  **(e)** 19″
**(9) (a)** 108 cm  **(b)** 233 cm  **(c)** $14.50  **(d)** 35.8°
  **(e)** 35.8°, 154 cm
**(11)** $\dfrac{12}{13}, \dfrac{5}{12}$  **(13)** 0.978″
**(15)** 50 mm, 259.8 mm, 326.8 mm, 271.9 mm
**(17) (a)** 250 mm  **(b)** 500 mm  **(c)** 683 mm
**(19) (a)** 41.4 mm  **(b)** 197.9 mm  **(21)** $\theta = 51.3°$, $\phi = 33.7°$

# CHAPTER 15   TRIGONOMETRY OF OBLIQUE TRIANGLES

## Self-Assessment Test No. 1 (Page 265)

**(1)** 0.7880  **(2)** −0.8462  **(3)** 51.7°, 128.3°
**(4)** 44.1°  **(5)** 129.7°

## Practice Exercise No. 1 (Page 266)

**(1)** 0.7071        **(3)** 0.5000        **(5)** 0.4695        **(7)** 0.0349
**(9)** 0.4741

## Practice Exercise No. 2 (Page 267)

**(1)** 52.2°, 127.8°        **(3)** 35°, 145°        **(5)** 54.1°, 125.9°
**(7)** 94.9°                **(9)** 43°, 137°

## Self-Assessment Test No. 2 (Page 268)

**(1)** 8.198°        **(2)** 118 mm        **(3)** 65.3° or 114.7°
**(4)** 72.5° or 107.5°

## Practice Exercise No. 3 (Page 271)

**(1)** 10.3        **(3)** 12.2        **(5)** 13.9        **(7)** 13.3

## Self-Assessment Test No. 3 (Page 272)

**(1)** $x = 10.85$ ft, $A = 86.8°$
**(2)** $A = 121.6°$, $B = 33.2°$, $C = 25.2°$

## Practice Exercise No. 4 (Page 274)

**(1)** 9.767        **(3)** 20.66        **(5)** 17.6°

## Practice Exercise No. 5 (Page 275)

**(1)** 15.21 cm²        **(3)** 17 216 mm²        **(5)** 100.6 in.²

## Practical Applications Exercises (Pages 275–278)

**(1)** $x = 136$ ft        **(3)** $a = 62$ m        **(5)** 17.4 cm        **(7)** 71.5 m

# APPENDIX A   SCIENTIFIC NOTATION

## Self-Assessment Test (Page 281)

**(1)** $2.56 \times 10^1$        **(2)** $2.11 \times 10^2$        **(3)** $2.1 \times 10^4$
**(4)** $1.976 \times 10^2$       **(5)** $1.01 \times 10^5$        **(6)** $7.52 \times 10^{-1}$
**(7)** $7.13 \times 10^{-3}$     **(8)** $3.7 \times 10^{-4}$      **(9)** $1.05 \times 10^{-2}$
**(10)** $1 \times 10^{-6}$       **(11)** 590                      **(12)** 9700
**(13)** 0.45                     **(14)** 0.000 13                 **(15)** 0.011

## Practice Exercise No. 1 (Page 282)

| | | |
|---|---|---|
| **(1)**  $7.77 \times 10^2$ | **(3)**  $4.2 \times 10^1$ | **(5)**  $6.693 \times 10^1$ |
| **(7)**  $9.33 \times 10^{-1}$ | **(9)**  $8.1349 \times 10^3$ | **(11)**  $7.78 \times 10^{-6}$ |
| **(13)**  $1.6992 \times 10^1$ | **(15)**  $8.9954 \times 10^3$ | **(17)**  $1.137 \times 10^3$ |
| **(19)**  $3.3 \times 10^{-4}$ | | |

## Practice Exercise No. 2 (Page 282)

| | | | |
|---|---|---|---|
| **(1)**  530 | **(3)**  51 | **(5)**  111 900 | **(7)**  18 000 |
| **(9)**  0.205 | **(11)**  0.0377 | **(13)**  0.000 016 3 | **(15)**  0.000 168 |

## Practice Exercise No. 3 (Page 283)

| | | |
|---|---|---|
| **(1)**  $1.8 \times 10^5$ | **(3)**  $4 \times 10^5$ | **(5)**  $6 \times 10^{-6}$ |
| **(7)**  $1 \times 10^{-1}$ | **(9)**  $2.376 \times 10^3$ | **(11)**  $4.226 \times 10^2$ |
| **(13)**  **(a)**  68 250 000 | **(b)**  0.000 435 1 | |

# APPENDIX B   METRICS

## Practice Exercise No. 1 (Page 286)

| | | | |
|---|---|---|---|
| **(1)**  13 000 | **(3)**  0.285 | **(5)**  27.051 | **(7)**  0.15 |

## Practice Exercise No. 2 (Page 287)

| | | |
|---|---|---|
| **(1)**  280 000 | **(3)**  8 715 100 | **(5)**  28 700 000 |

## Practice Exercise No. 3 (Page 287)

| | | |
|---|---|---|
| **(1)**  0.025 781 | **(3)**  12 000 | **(5)**  2 700 000 |

## Practice Exercise No. 4 (Pages 288–289)

| | | |
|---|---|---|
| **(1)**  0.287 | **(3)**  5000 | **(5)**  28 |

## Practice Exercise No. 5 (Page 289)

| | | | |
|---|---|---|---|
| **(1)**  1.250 | **(3)**  2.785 | **(5)**  2700 | **(7)**  2200 |

# APPENDIX C   THE CALCULATOR

## Practice Exercise No. 1 (Pages 292–293)

(1)   1183.984

(3)   15 234 668.1

(5)   1 491 954.674

(7)   2.028 913 96

(9)   499 627.777

## Practice Exercise No. 2 (Pages 294–295)

(1)   10.971 428 57

(3)   1.708 960 396

(5)   198.684 210 5

(7)   3.506 391 783

(9)   35.600 982 46

## Practice Exercise No. 3 (Page 297)

(1)   0.375

(3)   0.4286

(5)   2.875

(7)   33.3333

(9)   3

(11)   14

(13)   4

(15)   49

## Practice Exercise No. 4 (Page 298)

(1)   11.822

(3)   13.946

(5)   31.697

(7)   18

## Practice Exercise No. 5 (Page 299)

(1)   0.8771

(3)   0.4245

(5)   23.5°

(7)   12.05°

## Practice Exercise No. 6 (Page 301)

(1)   35.3°,  144.7°

(3)   177.1°

(5)   60.8°

(7)   80.8°,  99.2°

(9)   1.0°,  179.0°

# INDEX